Progress in

Sensory Physiology 4

Editors: H. Autrum D. Ottoson
E.R. Perl R.F. Schmidt H. Shimazu W.D. Willis

Editor-in-Chief: D. Ottoson

With Contributions by
N. Mei G. R. Martin A. Gallego

With 41 Figures

Springer-Verlag
Berlin Heidelberg New York Tokyo 1983

ISBN-13:978-3-642-69165-2 e-ISBN-13:978-3-642-69163-8
DOI: 10.1007/978-3-642-69163-8

Library of Congress Cataloging in Publication Data. Main entry under title: Progress in sensory physiology. Includes bibliographies and index. 1. Vision-Physiological aspects. 2. Senses and sensation. I. Autrum, Hansjochem. II. Ottoson, David, 1918–. III. Gouras, P. [DNLM: 1. Neurophysiology-Periodicals. 2. Sensation-Physiology-Periodicals. W1 PR681G] QP475.P89 612'.84 81-4430

2121/3140-543210

Contents

Sensory Structures in the Viscera

N. Mei*

C.N.R.S. - I.N.P.1, Département de Neurophysiologie Végétative,
31, Chemin Joseph-Aiguier, 13274 Marseille Cédex 2, France

* I particularly wish to thank Professor E. R. Perl for his helpful and interesting suggestions and for his very thorough correction of the manuscript for this review

1 Introduction

For a long time, particularly following Langley's work (see Gabella 1976), it was believed that the autonomic nervous system (vegetative or visceral) represented mainly or solely an efferent pathway. This view persisted even though early electrophysiological studies pointed out that visceral nerves contained a variety of afferent fibers. For instance, Adrian (1933), the pioneer in the single-fiber recording technique, described several different kinds of vagal afferent activity related to the cardiac and respiratory cycles. A few years later, Heymans and Neil (1937) discovered the carotid sinus baroreceptors whose discharge is related to the cardiac cycles. However, it took systematic histological analyses to provide quantitative information on the relative importance of the afferent side of visceral innervation. The work of Agostoni et al. (1957) in the cat and Evans and Murray (1954) in the rabbit represented principal contributions in this regard by showing that afferent fibers made up 80% of the vagal nerve. These observations were based upon light-microscopic comparison of the number of fibers in intact nerves with those in which efferent fibers had been eliminated by a bilateral transection central to the nodose ganglia. Even these light-microscopic histological studies indicated that a large proportion of unmyelinated fibers (approximately four-fifths) comprised the afferent vagal component. Light microscopy is only marginally useful for identifying unmyelinated fibers because of their very small size; recent data, in part electron-microscopic analyses, have confirmed the preponderance of afferent fibers in the vagal nerves, and showed that the unmylinated afferent component is even larger than earlier light-microscopic studies had suggested (Mei et al. 1980). Although the vagi do contain an exceptional number of afferent fibers, other visceral nerves similarly studied also proved to have a large afferent population.

Underestimating or overlooking the visceral afferent component is not due solely to morphological factors; it has functional reasons as well. First, the viscera are the source of relatively few and poorly localized sensations. There has been a tendency to emphasize pain as a visceral sensation and to overlook or disregard those physiological processes that are dependent upon visceral afferent input but are not appreciated at the conscious level. Division of major visceral nerves in man (vagotomy) or in animals does not result in easily demonstrable or severe dysfunction. This relative lack of effects is consistent with the automaticity of viscera such as the small intestine. Second, electrical stimulation of visceral afferent fibers often produces arterial hypertension and polypnea, reactions which are taken as indicative of pain; little or no attention is given to other, less dramatic, effects.

When proper attention is devoted to appropriate effector organs and to sensory experiences other than pain, afferent input from the viscera has been found to participate in a substantial number of neural, neurohumoral, and behavioral mechanisms. This review proposes to concentrate on the morphological, electrophysiological, and other functional features of visceral afferent neurons that have come to light over the past two decades in the peripheral autonomic nerves. It is not a complete catalogue of the literature, but rather an attempt to illustrate, with selected examples, the major current theories and methods of investigation. Some of the most interesting recent studies have taken place on the neural apparatus associated with the digestive system, and as a consequence it will receive considerable attention.[1]

2 Methodological Considerations

The methods available for studying visceral afferent arrangements are no different from those used in the somatic system; however, there are special problems due to the fact that the peripheral visceral nervous system is composed almost exclusively or extremely small diameter nerve fibers. This means that some of the techniques readily applicable to somatic afferent arrangements have limited applicability or are more difficult to use. Histological analysis requires the use of an electron microscope to study the large number of nonmedullated fibers, many of which have cross-sectional diameters at the limit of light optical resolution. To determine the number of fibers in the nerve and the relative proportion of different sizes or classes, a sample of 5% – 10% of the nerve's cross-sectional area is required (Jammes et al. 1981). Even with the aid of computerized data processing, electron-microscopic samples of this dimension require a prolonged and painstaking analysis.

The application of horseradish peroxidase (HRP) as a tracer of pathways to the visceral afferent supply has been an important recent achievement. This procedure has been used to identify cell bodies of neurons with processes reaching to the periphery by the retrograde transport of the peroxidase from an injection point in the viscera or the digestive tube (Neuhuber and Niederle 1979) or from an autonomic nerve such as the aortic nerve (Portalier and Vigier 1979). The retrograde transport of HRP has also been used to assess the relative contributions of fibers supplying a given organ of the digestive tube from different nerves, such as the vagus and the splanchnic (El Ouazzani 1981). Less frequently, HRP has been injected into neuronal somata to be transported in the anterograde fashion so that the diameter of the axon can be determined (Gallego and Eyzaguirre 1978).

1 The exceptions, of course, are the well-established functions of the baro- and chemoreceptors of the carotid sinus and aortic regions and the bronchopulmonary afferent fibers in cardiovascular and respiratory regulation

Recording the compound action potential of a visceral nerve has proven a useful electrophysiological estimation of activity associated with fibers of different conduction velocities and, hence, different diameters, a procedure used in Heinbecker and O'Leary's study (1933) on the vagus. However, this method has serious limitations for the analysis of functional features. First, it lends itself only to population studies and cannot give information about individual nerve fiber. Second, there is no way to discriminate between afferent and efferent fibers populations in intact nerves; only if one or the other population is removed by chronic section or destruction of the motor supply, for example, would it be possible to define the afferent components.

On the other hand, the electrophysiological approach at the unitary level has provided a great deal of information about the receptive properties and the characteristics of the nerve fibers of visceral primary afferent neurons. Consequently, much of the current thinking about visceral afferent mechanisms is derived from the observation of unitary activity. The unitary studies have been mainly made using two approaches: (a) the isolation of activity to an identifiable fiber by cutting down a nerve (Fig. 1A), which was first employed by Adrian (1933) in the visceral domain, and (b) the use of microelectrodes inserted into the sensory ganglia to record electrical activity extracellularly (Fig. 1B). The latter is par-

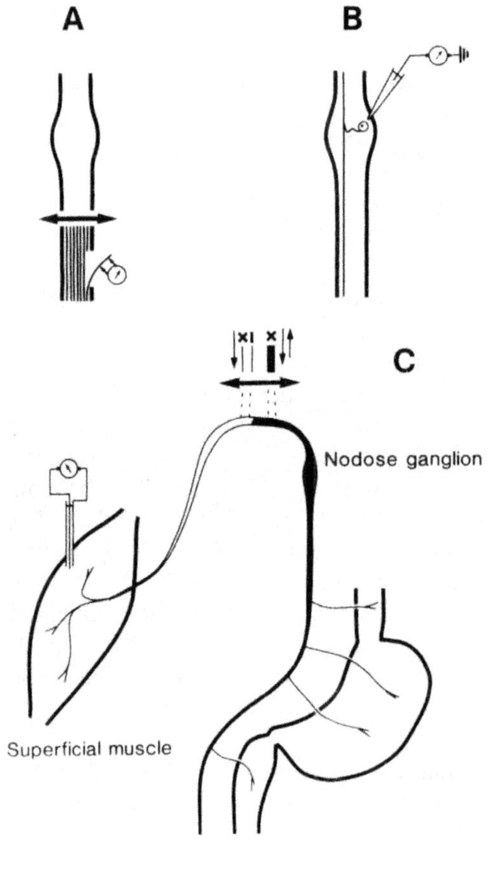

Fig. 1A – C. Microelectrophysiological techniques used to study the visceral afferents (vagal nerve with nodose ganglion is represented here). **A** Isolated nerve technique; **B** microelectrode technique; **C** crossed nervous suture technique. Vagal afferents are recorded by means of coaxial electrodes implanted in superficial muscle. *X*, vagus nerve; *XI*, spinal nerve

ticularly well suited for the study of neurons with unmyelinated afferent fibers (Mei 1970a, 1978). A third method of studying individual neurons deserves attention because of its possibilities for chronic study (Fig. 1C); it is based upon the process of innervation of a skeletal muscle motor nerve by central branches of sensory cells (Rousseau 1970; Falempin 1981).

To adequately study the functional features of afferent fibers it is necessary to sample appropriate stimuli and visceral effector activity. Modern approaches include recording muscle contraction through pressure variations sampled by intraluminal balloons, open-tipped catheters or intraluminal microballoons, and electromyographic recording, which for the intestinal system is usually monitored by use of suction electrodes (see Gonella and Gardette 1974). Recording effector-directed neuronal activity is more difficult. Discharges from motoneurons, preganglionic or postganglionic neurons, can be recorded by splitting motor nerves into fine filaments so that single unitary activity can be identified. It is also possible to utilize the crossed innervation technique mentioned in the previous paragraph; in this technique denervated somatic muscles are innervated from visceral sources, an approach which has been used for study of digestive motor activity (see Roman and Gonella 1981). Finally, secretory output can be determined utilizing radioimmunoassay whenever the substances released are immunologically active, an approach that is remarkably sensitive and precise for detecting hormones such as those of the intestine and insulin.

3 Morphological Considerations

3.1 Peripheral Visceral Afferent Pathways

Standard textbook descriptions point out that visceral afferent impulses travel over a great variety of peripheral pathways. Afferent fibers are found in both sympathetic and parasympathetic systems. For purposes of discussion, I propose defining afferent fibers according to the division of the autonomic system through which they pass and labeling them accordingly "sympathetic" and "parasympathetic" afferent fibers (Fig. 2). Admittedly, this convention breaks with the tradition of using sympathetic and parasympathetic to designate efferent components only, but it has been utilized by other authors.[1]

In visceral nerves, afferent fibers generally appear intermingled with efferent fibers; however, near the juncture with the cerebrospinal axis a separation of these fibers takes place. Afferent fibers of the sympathetic system and the lumbosacral parasympathetic system, which are separate from the efferent fibers and generally enter the spinal cord in the dorsal roots or at the medullary level, form essentially pure sensory nerves, as in the case of Hering's and Cyon-Ludwig's nerves. In some cranial parasympathetic nerves, afferent and efferent fibers stay

1 Another term, "general visceral afferents" (Crosby et al. 1962), is less desirable because it is not sufficiently specific about the pathway. Both sympathetic and parasympathetic afferent fibers have their cell bodies in spinal or cranial ganglia and, therefore, do not differ in this respect from their somatic counterparts

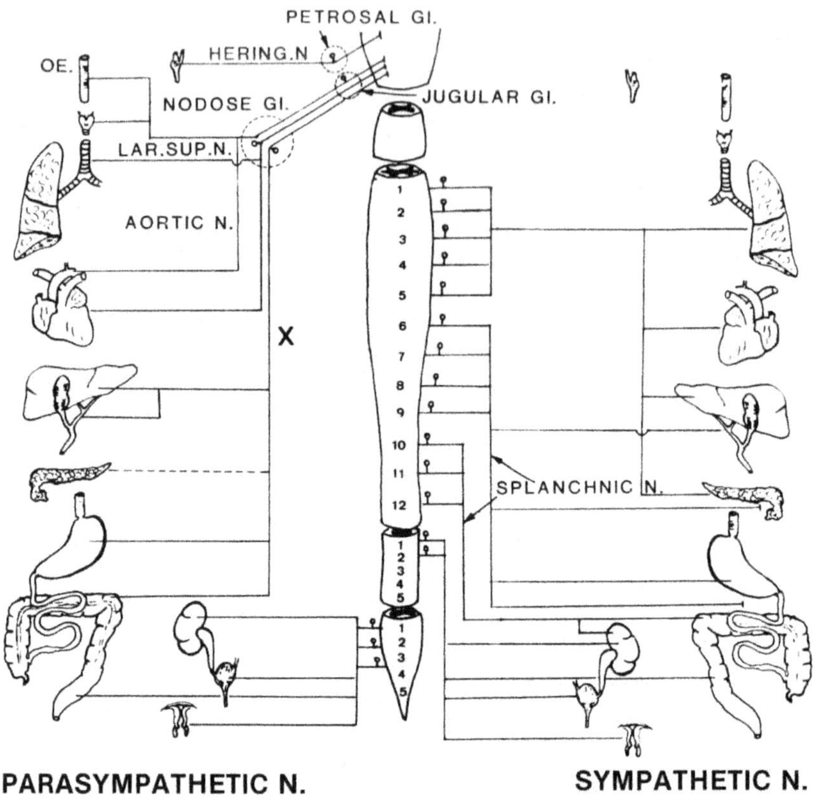

Fig. 2. Sympathetic and parasympathetic pathways of visceral afferent fibers. *OE,* cervical esophagus; *N,* nerve; *GL,* ganglion; *LAR. SUP. N.*, laryngeal superior nerve; *X,* vagus nerve

within the same nerve bundle until entry into the central nervous system. Recently it has been shown that there are a number of special cases as well. There are three examples: (a) In the vagus nerve an almost complete separation of efferent (motor) and afferent fibers takes place at the nodose ganglion with the motor fibers running in anatomically distinct ventral bundles (Mei 1966; Darcy et al. 1979), and a nominal separation of these fibers occurs at the bulbar roots (Paintal 1963). (b) Some afferent fibers reach medullary levels through somatic nerves, such as certain of those from the cardiac region that pass by way of the phrenic nerves (Ruckebusch 1961). (c) A large number of afferent fibers in the pelvic region with cell bodies in the spinal dorsal root ganglia have central branches which reach the spinal cord through ventral roots (Coggeshall et al. 1974).

Neurons other than those of the dorsal root and cranial ganglia can serve afferent functions. Neurons of this type would have a receptive terminal in the visceral organ or structure and a central terminal synapsing within the organ itself or in nearby peripheral ganglia. Afferent connections of this type have been shown to exist for the intestine, where short reflexes organized entirely outside of the central nervous system play an important part (Kreulen and Szurszewski 1979; Szurszewski 1981).

The afferent nervous supply to visceral structures may pass by way of sympathetic nerves alone, parasympathetic nerves alone, or both systems. The intestine, for example, receives most of its afferent innervation from the vagus, whereas its supporting mesentery is almost entirely innervated by fibers that pass through sympathetic nerves. On the other hand, the lower esophagus and the stomach have afferent terminals whose centrally directed fibers pass through both sympathetic and parasympathetic nerves. It should be noted that even when the afferent fibers to an organ pass through a nerve of one system, more than one nerve may be involved; a case in point is the upper portion of the thoracic esophagus whose afferent supply passes by way of both the vagus and the superior laryngeal nerves.

3.2 Afferent Composition of Visceral Nerves

The most important contribution of recent research on visceral afferent mechanisms has demonstrated a prominence of afferent fibers in both parasympathetic and sympathetic visceral nerves. Afferent fibers make up 80% of the 30000 (light-microscopic level) to 50000 (electron-microscopic level) fibers of the vagus nerve (Agostoni et al. 1957; Mei et al. 1980). As already mentioned, underestimation of the magnitude of the sensory component of visceral nerves can be explained by the fact that the visceral afferent fibers are predominantly unmyelinated and, therefore, difficult to visualize by light microscopy.

The cross-sectional diameter of the unmyelinated fibers varies between 0.4 and 3 μm, with an average of approximately 1 μm. Visceral afferent fibers also have myelinated fibers of varying sizes from 2 μm up to 12 μm. Although there is evidence that B fibers exist in the afferent component of autonomic nerves on the grounds of conduction velocity and excitability cycle (see Paintal 1963; Mei 1970a, and Fig. 7), the visceral afferent fibers are conventionally classified in Aβ, Aγ, and Aδ groups. The term B fiber is generally reserved for preganglionic autonomic fibers, according to the Gasser and Erlanger classification.

The range of diameters for myelinated and unmyelinated fibers suggests that the two populations overlap. This has been established to exist in the vagal nerve branches innervating the lungs and making up the abdominal distribution (Mei et al. 1980; Jammes et al. 1981).

The afferent component of visceral nerves varies in the proportions of fibers of different sizes although, in general, both myelinated and unmyelinated types are present in main trunks as well as in peripheral branches (Mei 1981). In the vagus, the largest diameter afferent fibers are in the cardiovascular neurons, whereas in the splanchnic nerves the largest diameter fibers are those associated with the pacinian corpuscles of the mesentery. There is a special pattern in the distribution to the digestive system: the proportion of unmyelinated fibers increases from the oral to the aboral end, so that at the intestinal level the afferent fibers are almost entirely unmyelinated.

3.3 Features of Visceral Primary Afferent Neurons

Variation in the diameter of primary afferent fibers in the peripheral visceral nerves is the structural variation among the primary afferent neurons most frequently discussed. However, noteworthy information has accumulated on other morphological aspects that deserves mention.

3.3.1 Differences Along the Course of Primary Afferent Fibers

Van Gehuchten (1892a, b) and Cajal (1909) proposed the long-held view that the peripherally directed process of a primary afferent neuron corresponds to the dendrite of the more typical nerve cell, and the fiber that passes into the central nervous system is the equivalent of the axon. This proposal was based upon embryological and phylogenetic considerations, as well as on the direction in which action potentials propagate under physiological conditions. The spinal ganglia containing the cell bodies of primary afferent neurons in their derivation from the neuronal crest initially have peripheral locations; the neurons they contain are bipolar early in embryonic life, becoming unipolar later in the higher vertebrates. Differences in the ultrastructure of the peripheral and the centrally directed fibers, such as the density of microtubules related to protein transport, have emerged (Droz 1965; Lasek 1968). In spite of such differences, the current tendency is to refer to both processes as axons because of their length (cf. Lieberman 1976). (Etymologically, the word "dendrite" derived from the Greek *dendron* [tree] and should be applied only to the short ramifications of processes emanating from the soma.) This nonselective use of the term axon has been extended to visceral primary afferent neurons some of which, like the smaller cells of spinal ganglia supplying somatic structures, have central fibers that are much shorter than those passing peripherally.

In broad terms, the central process of a primary afferent neuron has a smaller diameter than the peripheral one, regardless of whether the fibers are myelinated or nonmyelinated (Fig. 3). This observation initially made at the light-micro-

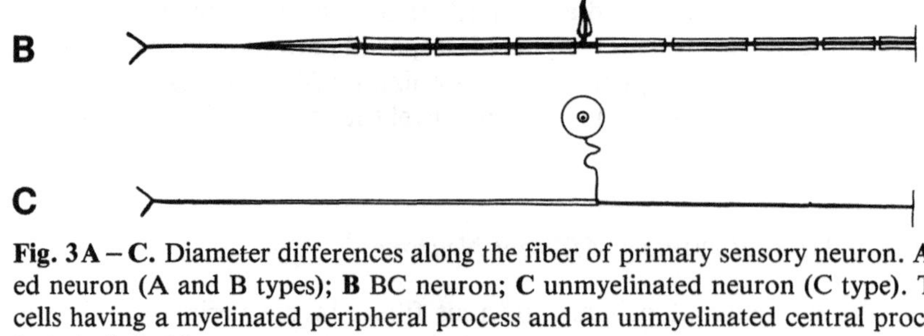

Fig. 3A – C. Diameter differences along the fiber of primary sensory neuron. **A** Myelinated neuron (A and B types); **B** BC neuron; **C** unmyelinated neuron (C type). The case of cells having a myelinated peripheral process and an unmyelinated central process has not been included here

scopic level by Cajal (1909) and Ranson et al. (1933) has recently been confirmed by systematic electrophysiological and electron-microscopic analyses (Gasser 1955; Mei 1970a; Mei et al. 1980). This difference in diameter between peripheral and central fibers is characteristic throughout the length of the neuraxis.

There are other indications of heterogeneity along the length of primary afferent fibers. Certain small diameter fibers, medullated in peripheral nerves, are unmyelinated in the central part of their trajectory. A significant number of fibers that are unmyelinated distally in their peripheral course, acquire myelin prior to reaching the level of the cell's soma (Duclaux et al. 1976), as diagrammed in Fig. 3 (mixt fibers, type BC). The number of unmyelinated fibers that are associated with a Schwann cell is statistically higher central to the spinal ganglion than peripherally, an observation indicating reorganization of some type from one part of the pathway to another.

3.3.2 The Cell Body and the Stem of Its Process

The somas of visceral primary afferent neurons are not known to differ in any significant way from those of their somatic counterparts (see Lieberman 1976). The stem of the process is ordinarily easily distinguished by its conical shape, but in the absence of a distinct cone can still be discerned by a scarcity of Nissl bodies, and parallel arrays of numerous microfilaments and microtubules. Initially the stem is unmyelinated and generally exhibits convolutions of varying complexity, forming a glomerulus in the vicinity of or actually enveloping the soma (Cajal 1909; Lieberman 1976). The stem bifurcates to form the centrally and peripherally directed fibers at a distance of from a few tenths to several hundred μm from the soma (Ha 1970).

The presence and nature of some structural arrangements are still controversial. Uncertainty still exists about sympathetic innervation in the vicinity of the afferent cell bodies and the glomerular arrangements near them. An efferent innervation was described by Cajal; however, definite identification has not yet been made at the ultrastructural level (Lieberman 1976; Mei, unpublished). Similarly, atypical multipolar cells, frequently reported in the literature (Lieberman 1976), have not been satisfactorily documented to be considered other than exceptional or unexplained formations. The same uncertainty exists for serotonergic neurons evidenced in the nodose ganglion (Gaudin-Chazal et al. 1981).

The location of cell bodies within the afferent ganglia varies. In spinal ganglia, the cells tend to be located peripherally, partially or completely surrounding the bundle of fibers, whereas in the nodose ganglion of the vagus, the neuronal somas form isolated clusters surrounded by strands of nerve fibers. The cell bodies are not randomly distributed, rather being grouped according to the distribution of their peripheral terminals. This topical arrangement has been established by a variety of techniques including retrograde degeneration, histochemical determination of peripherally injected peroxidase, autoradiographic localization of tritium-labeled proline, and electrophysiological recording of unitary activity. These studies have shown that the nodose ganglion contains pulmonary, cardiovascular, aortic, and gastrointestinal regions (Molhant 1913; Mei 1970a; Portalier and Vigier 1979; Garcia et al. 1979). A similar topographical organization has been reported for the gasserian ganglion (Beaudreau and Jerge 1968; Roubien and Mei 1970) and spinal ganglia (Arlhac 1972; Burton and McFarlane 1973). The topographical arrangement within the ganglia is possibly related to a somatotopic arrangement in the dorsal roots (Burton and McFarlane 1973) and the roots of cranial nerves (Paintal 1963), as well as more centrally. A long-standing dictum in neurohistological textbooks, that the caliber of a nerve fiber is directly related to the diameter of the perikaryon, has been called into question. It has been specifically shown in the nodose ganglion that unmyelinated fibers may possess somas of a size comparable to those associated with myelinated fibers (Gallego and Eyzaguirre 1978). Nevertheless, a more precise correspondence between cell body size and fiber caliber seems to exist for neurons of the spinal cord (Yoshida and Matsuda 1979).

3.3.3 The Peripheral Receptive Structures

The receptive arrangements for afferent units in the autonomic distribution are termed visceral receptors or, alternatively, enteroreceptors. Most available morphological information comes from light-microscopic observations, since the extensive distribution of their afferent terminals without clearly discernible capsu-

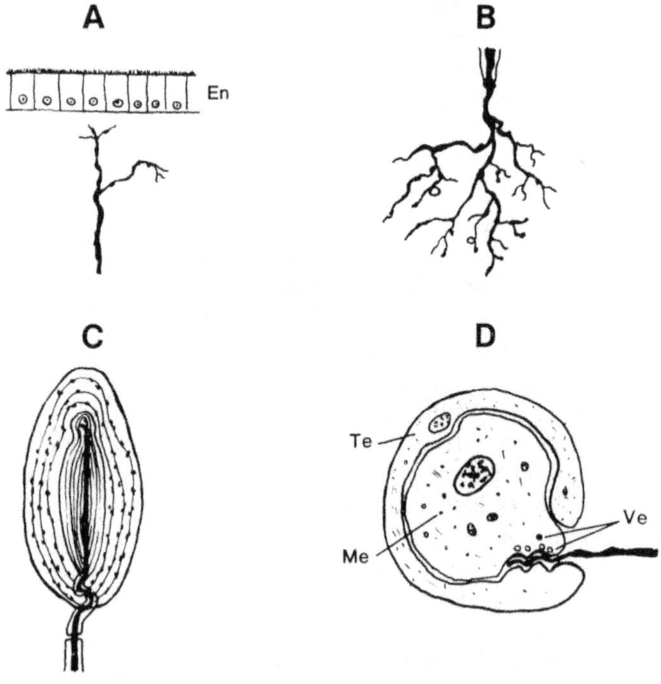

Fig. 4A–D. Main kinds of visceral sensory endings. **A** Simple branching ending of small intestine; *En*, enterocytes. **B** Rather more complex ending of cardiovascular system; **C** pacinian corpuscle of mesentery; **D** arterial chemoreceptor. *Me*, metabolic cell; *Te*, tentacular cell; *Ve*, synaptic vesicles

lation, granules, or associated cells often makes it difficult to distinguish them from efferent endings at the electron-microscopic level. This problem may be rectified in the future should a technique for separately marking afferent fibers and their terminals be developed.

From what is now known, the structure of enteroreceptors in general seems to be uncomplicated. The afferent fiber, medullated or unmedullated, ends as simple terminations in the viscera after a greater or lesser degree of branching (Fig 4A). The receptive terminal, or the "receptor", as it is often called, never appears to be myelinated, implying that medullated fibers lose their myelin sheath at some point in their distal path. Enteroreceptors principally differ from each other in the location of their terminals and in the number of branches or subdivisions of their afferent fibers. Most digestive and respiratory receptors have relatively few terminal branches, whereas cardiovascular receptors tend to have many terminal branches and relatively complex termination patterns for each fiber (Fig. 4B).

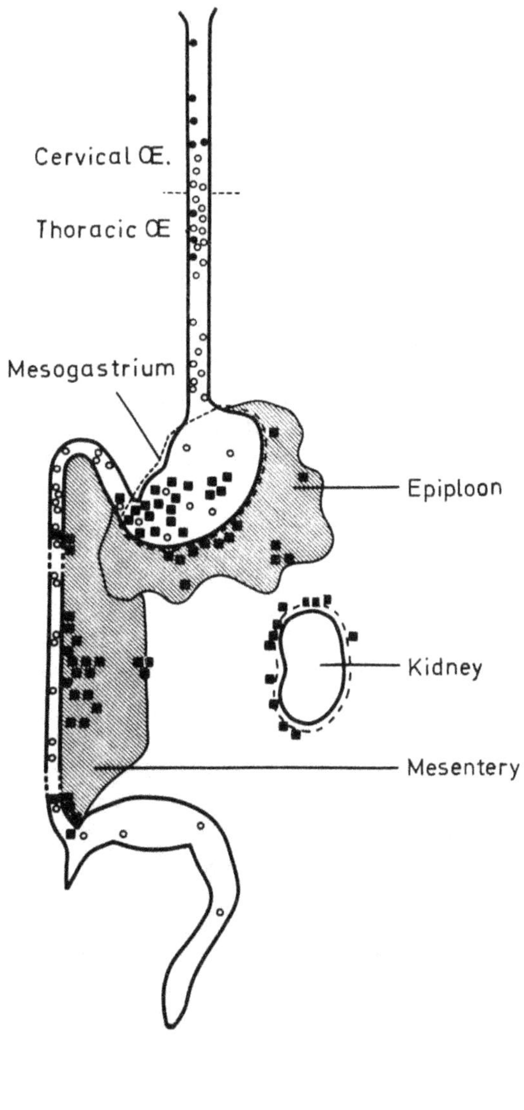

Fig. 5. Distribution of mechanoreceptors in the digestive tract. These results were obtained in anesthetized cats with the microelectrode technique. Each symbol (circle, dot, or square) represents one receptor.
■ Splanchnic mechanoreceptors;
○ vagal mechanoreceptors;
● laryngeal mechanoreceptors

The distal branches do not exhibit a uniform pattern; they can be very sinuous or fairly straight, and their diameter may remain constant or become irregular with numerous swellings, the usual case for intestinal afferent fibers (Mei 1978). It appears a general rule that the fiber's Schwann cell covering turns back upon itself, leaving the termination branches free of a covering. As ultrastructural investigation of these terminals progresses, it may develop that differences in "specific" types of endings are dispersed in different tissues, as has been established in certain of the "free nerve endings" in somatic tissues.

Some specializations for visceral receptors have already been noted. One example is the pseudo spindle receptors reported for the myocardium and endocardium (Khabarova 1963). Certain esophageal afferent terminals that are characterized by an inconspicuous capsule represent another instance of specialization (Temesrekasi 1956). Of course, the mesenteric pacinian corpuscles in the cat, which are identical to the highly organized equivalent structures in somatic regions, are still another case (see Fig. 4C). The well-verified and extensively studied arterial chemoreceptors represent still another highly organized afferent arrangement. The chemoreceptors of the carotid body and aortic region consist of two cells, one centrally located with a large number of inclusions and a sizable nucleus (the metabolic cell), and the other more peripherally positioned and displaying a large number of infoldings (the tentacular cell). At the ultrastructural level, nerve terminals, presumably of afferent fibers, are interdigitated between the metabolic and the tentacular cells (Howe and Neil 1972), as diagrammed in Fig. 4D.

Histological and electrophysiological observations over the past two decades emphasize the ubiquitous distribution of afferent terminals in the viscera. Apparently, all visceral structures have sensory innervation of one type or another. Moreover, afferent terminals are present in the various parts of a viscus: mucosa, muscle, or serosa. On the other hand, visceral receptors are not uniformly dense throughout a viscus any more than they are in a somatic tissue or organ. A clear example in the gastrointestinal tract is the concentration of afferent terminals in the sphincter regions as compared with other parts of the intestinal tube (Fig. 5).

4 Electrophysiological Studies

Under physiological conditions, once an afferent impulse is generated at the receptive terminal of a primary neuron, it travels unchanged to the region of its first synaptic junction at its central terminal. Thus, sampling the impulse activity along the fiber, at the ganglion cell, or even by the responses of an effector organ (as in crossed innervation experiments) will yield the same information. Most of our present knowledge on physiological characteristics of visceral afferent structures comes from such recordings of impulse traffic at one or another point along the primary afferent neuron.

4.1 General Physiological Properties of Visceral Primary Afferent Neurons

4.1.1 Receptive Mechanisms

There is no evidence in the literature that the initiation of receptor or generator potentials and of action potentials is different for enteroreceptors than it is for those sense organs located in somatic structures. It is true that enteroreceptors generally show a high level of sensitivity to local anesthetics and to various pharmacologically active substances such as veratrine, bradykinin, and phenyl diguanide. This special responsiveness possibly can be explained by the fact that most enteroreceptors have unmyelinated or thinly myelinated fibers which may be more accessible to chemical agents than the more thickly myelinated fibers of somatic nerves. Furthermore, the terminal region where the conducted action potential is generated may possess special sensitivity to chemical agents (Paintal 1968). As a consequence of the responsiveness to such pharmacologically active agents, it has been possible to study several aspects of visceral afferent mechanisms through injection of agents into the general systemic circulation or the local circulation of particular organs (Paintal 1972).

In the fashion traditional for sensory receptors, enteroreceptors are often divided into two groups: slowly adapting units, which, incidently, often exhibit a striking dynamic response, and rapidly adapting receptors, some of which give "on" and "off" responses. Although these two broad categories have been used to classify visceral receptors, it is now apparent that the variety of stimuli selectively activating one or another visceral receptor is substantially greater than had been previously thought. This selectivity exists either because of the terminal's location in particular tissues or specificity in the terminal characteristics. Finally, it should be emphasized that modulation of response may take place at the visceral receptor itself. Three mechanisms could be involved in the latter proposal: (a) Sympathetic efferent fibers may directly alter afferent receptor activity, as has been suggested for the pacinian corpuscle (Akoev et al. 1976). (b) Efferent fibers may act on an effector organ, as do the gamma fibers on the intrafusal muscle fibers of the muscle spindle, and modify mechanical factors; this mechanism has been suggested for the sympathetic fibers innervating the carotid sinus (Sampson and Mills 1970). (c) The efferent fibers may modulate the local circulation around different terminals, a mechanism proposed for some endings in epithelial structures and also for the pacinian corpuscle (Leitner and Perl 1964; Freeman and Rowe 1981).

4.1.2 Impulse Conduction in the Sensory Ganglion

Potential changes reflecting nerve impulses can be recorded in the cell body of primary afferent neurons even though these cell bodies are located on a side branch of the main fiber passing from the periphery to the central nervous system. This applies to visceral afferent neurons as well as those of the somatic system, as has been established by extracellular and intracellular recordings from ganglion cells themselves (Svaetichin 1951, 1957; Ito 1959; Sato and Austin 1961; Bessou et al. 1971; Jaffe and Sampson 1976; Gallego and Eyzaguirre 1978; Mat-

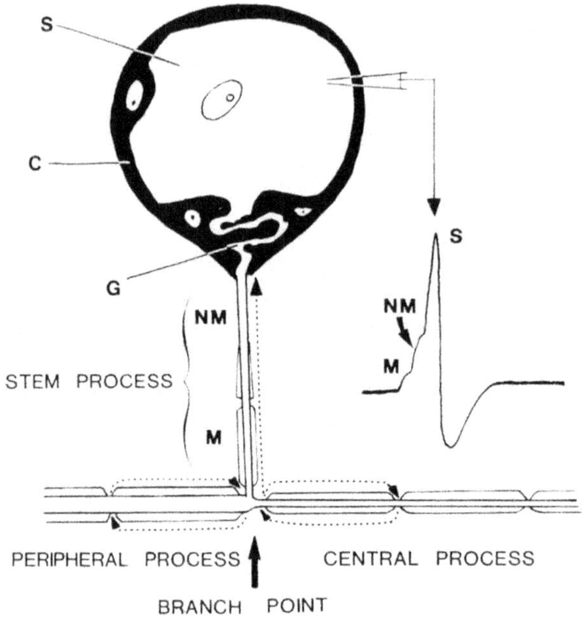

Fig. 6. Influx propagation in ganglion. At the branch point, an action potential depolarizes both the central process and the stem process. In the last part, the propagation involving an electrotonic mechanism may be detected in the impaled soma (*S*) with an intracellular electrode. *M*, myelinated part of stem process; *NM*, nonmyelinated part of stem process; *C*, capsule; *G*, glomerulus

suda et al. 1978; Yoshida and Matsuda 1979). Since the soma is connected to the peripherally and centrally directed fibers in most ganglion cells by a T-shaped process, a question has arisen as to whether the impulse actively invades this process and the cell body or whether the activity recorded at the level of the cell body represents electrotonic spread from an active process in the fibers. Arguments have been proposed in favor of electrotonic spread of activity for both myelinated and unmyelinated afferent fibers (Fig. 6). In any case, the likelihood of simultaneous activation of both the process connecting the soma to the fiber and its central continuation by the impulses arriving from the periphery is favored by two factors: (a) the central extension of a fiber typically has a smaller diameter than the peripheral one so that a smaller area of membrane needs to be depolarized, and (b) the field potentials from neighboring and simultaneously active neurons, particularly those having unmyelinated fibers, since their relatively long-duration action potentials aid in depolarizing the unexcited parts of the fiber (Bessou et al. 1971; Gallego and Eyzaguirre 1978). As an aside, it might be noted that the longer duration and the greater magnitude of action potentials from unmyelinated fiber cells probably is a factor in facilitating simultaneous recording from several primary afferent units with unmyelinated fibers in afferent ganglia, since the effects of tissue capacitance in shunting the electronic spread would be lessened.

4.2 Functional Properties of Visceral Afferent Neurons

Visceral afferent units can be classified into three major groups, according to whether their principal activity is associated with mechanical stimuli, chemical

Table 1. Receptors in the digestive tract and its annexes

Receptors	Location	Fiber pathway	Stimuli	Main contributors
Muscular mechano-receptors				
1	Esophagus	Vagus, superior laryngeal	Contraction, distention	Andrews (1955, 1957), Mei (1970b), Clarke and Davison (1975), Falempin et al. (1978)
2	Gastrointestinal tract	Vagus	Contraction, distention	Paintal (1954), Iggo (1955, 1957a), Leek (1969), Mei (1970b), Andrews et al. (1980)
3	Gastrointestinal tract	Splanchnic	Contraction, distention	Morrisson (1973), Ranieri et al. (1973), Floyd et al. (1976)
4	Colon	Vagus	Contraction, distention	Mei (1970b)
5	Colon, rectum	Pelvic	Distention	Clifton et al. (1976)
6	Gall-bladder	Vagus	Distention, contraction	Crousillat and Ranieri (1980)
Mucosal mechano-receptors[a]				
1	Esophagus, gastrointestinal tract	Vagus	Mucosal stroking, strong distention, contraction, HCl	Paintal (1957), Harding and Leek (1972a, b), Davison (1972), Clarke and Davison (1978)
2	Colon, rectum	Pelvic	Mucosal pinching	Clifton et al. (1976)[b]
Serosal mechano-receptors				
1 (Free endings)	Mesentery (intestine, bladder)	Splanchnic, hypogastric	Movement of mesentery, strong distention of viscus	Bessou and Perl (1966), Morrisson (1973), Ranieri et al. (1973), Floyd et al. (1976)
2 (Free endings)	Mesentery (lower esophageal sphincter, pylorus)	Vagus	Movement of mesentery, strong distention of viscus	Mei (1970b), Clerc and Mei (1981)
3 (Free endings)	Mesentery (gall-bladder)	Splanchnic	Movement of mesentery, strong distention of viscus	Crousillat and Ranieri (1980)

Table 1 (continued)

Receptors	Location	Fiber pathway	Stimuli	Main contributors
Serosal mechano-receptors				
4 Pacinian cor-puscles	Mesentery	Splanchnic	Movement of mesentery, strong distention of viscus	Gammon and Bronck (1935), Ranieri et al. (1976)
Chemo-receptors				
Alkalio and acid receptors	Gastrointestinal mucosa	Vagus	pH, alkali, acid	Iggo (1957b), Andrews and Andrews (1971), El Ouazzani and Mei (1981)
Gluco-receptors				
1	Gastric mucosa	Vagus	Glucose, other carbohydrates	El Ouazzani and Mei (1981)
2	Intestinal mucosa	Vagus	Glucose, other carbohydrates	Mei (1978), El Ouazzani and Mei (1981)
3	Intestinal mucosa	Splanchnic	Glucose, other carbohydrates	Sharma and Nasset (1962), Hardcastle et al. (1978), Perrin et al. (1981)
Amino acid receptors	Intestinal mucosa	Vagus	Amino acids	Jeanningros and Mei (1980), Jeanningros (1982)
Thermo-receptors				
1	Serous membrane (dorsal abdominal wall)	Splanchnic	Warmth	Riedel (1976)
2	Esophageal, gastrointestinal mucosa	Vagus	Cold	Gupta et al. (1979), El Ouazzani and Mei (1979)
3	Esophageal, gastrointestinal mucosa	Vagus	Warmth	
4	Esophageal, gastrointestinal mucosa	Vagus	Warmth, cold	El Ouazzani and Mei (1979, 1982)

[a] Mucosal mechanoreceptors involve both true mechanoreceptors and multimodal receptors

[b] Concerns neurons the fibers of which follow the spinal ventral roots; their pathway in the pelvic nerves is purely putative

changes, or thermal stimuli. The comments that follow in this section are intended to provide information on certain common characteristics of each group. For more detailed and complete discussions on the responsive characteristics, the reader is referred to Paintal (1963, 1973) for a discussion of enteroreceptors, to Leek (1972, 1977) for receptors associated with the digestive tract, to Mei (1978) for a consideration of digestive chemoreceptors, and to El Ouazzani and Mei (1979, 1981) for thermoreceptors of the digestive tract. As an orientation and a summary, Table 1 provides current information on the principal types of enteroreceptors within the digestive tract.

4.2.1. Mechanoreceptors

Classically, mechanoreceptors have been divided into rapidly adapting and slowly adapting types; however, this division may be based upon inappropriate

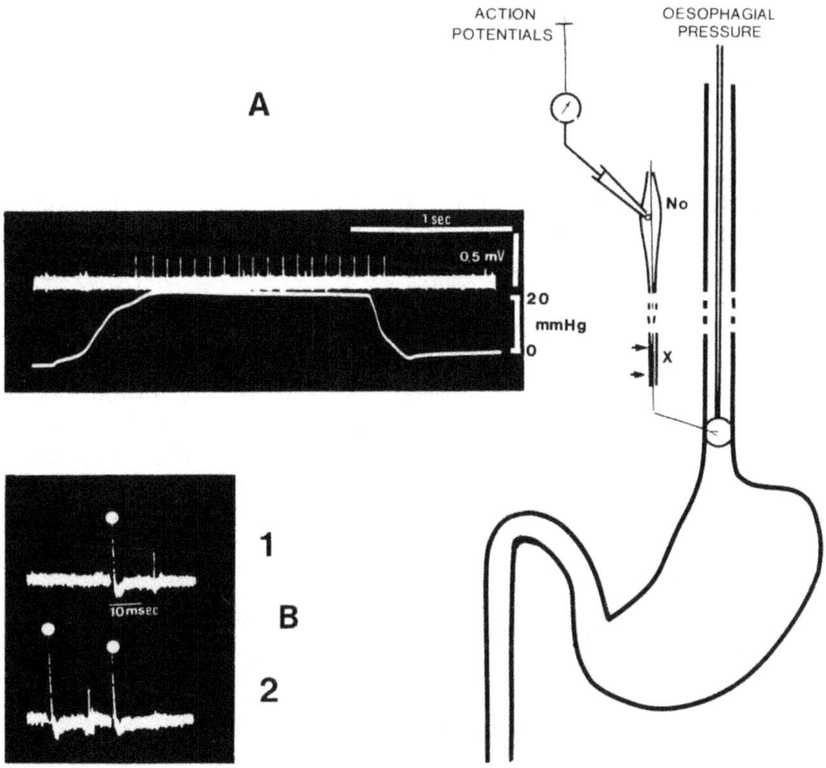

Fig. 7A, B. Slowly adapting receptor located in the esophageal muscular layer. In this figure, as in the following ones, sensory units were recorded with the microelectrode technique. **A** Response elicited by esophageal distension; **B** single response produced by electrical stimulation of vagus (*dots*). *1* single stimulation; from this, the conduction velocity was calculated (V = 7 m/sec). *2* double stimulation. The neuron was not excitable twice when the second stimulus occurred between 0 and 60 msec (subnormal phase). It was concluded that the fiber belonged to the B type. *No,* nodose ganglion; *X*, vagus nerve; *arrows*, electrical stimulation

criteria (Burgess and Perl 1973). Given that displacement itself is an appropriate stimulus for a mechanoreceptor, afferent units that adapt very slowly to a maintained position are distinguishable from those adapting rapidly. It should be noted that for many of the so-called rapidly adapting mechanoreceptors, the appropriate or "adequate" stimulus is actually movement or a derivative of movement, such as acceleration; when such rapidly adapting mechanoreceptors are tested using constant velocity or a constant derivative of velocity, discharges continue for the duration of the stimulus. These distinctions have been more thoroughly analyzed for the somatic system (e.g., Brown and Iggo 1967; Burgess and Perl 1973).

4.2.1.1 Slowly Adapting. Mechanoreceptors that adapt slowly have been identified and studied in the respiratory system (Widdicombe 1954a, b; Fillenz and Widdicombe 1972), in cardiovascular structures (Paintal 1972, 1977), in the urinary bladder (Iggo 1955, Arlhac 1972, Floyd et al. 1976, 1977), and in the kidney (Aström and Crafoord 1968; Beecham and Kunze 1969; Niijima 1971). The kinds of stimuli that activate the slowly adapting visceral mechanoreceptors can be produced by a variety of events such as distention or stretching of the innervated organ, as well as by intrinsically produced contraction. These events may be part of normal physiological activity, as in the gastrointestinal tract where they are produced by the process of normal motility. Because distention or contraction can alter the tension of the walls of the hollow viscus, the sense organs therein have been labeled "tension receptors" (Leek 1977). This population of slowly adapting mechanoreceptors has low thresholds for mechanical stimuli, and many of them are spontaneously active under normal physiological circumstances, as is illustrated in Fig. 8. Generally, when a viscus is innervated by both myelinated and unmyelinated afferent fibers, as in the cardiovascular and respiratory systems, the slowly adapting mechanoreceptors tend to have myelinated afferent fibers. However, even though the principal innervation to a viscus may stem from unmyelinated fibers (the gastrointestinal tract, for example), slowly adapting receptors may still be present. The differences that have been described in the literature for slowly adapting mechanoreceptors with myelinated and unmyelinated fibers seem to be principally related to inherent characteristics of

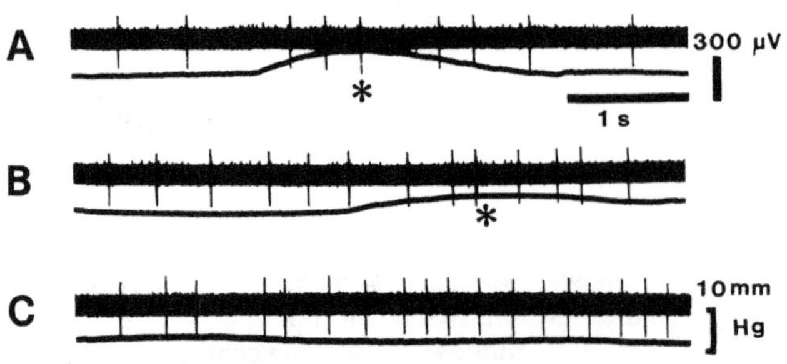

Fig. 8A – C. Digestive muscular receptor located in the lower esophageal sphincter. Effect of spontaneous contraction. **A**, **B**, and **C** are continuous. After 2 visible contractions (*) in **A** and **B**, the sphincter presented a tonic closure which elicited a sustained discharge in the receptor

the parent fiber; for instance, those with myelinated fibers have substantially higher frequencies of discharge (100 – 300 impulses/second) than those with unmyelinated fibers (10 – 30 impulses/second). The slowly adapting mechanoreceptors of the viscera are largely located within the muscular layers, being much less frequent in the serous membrane. The intramuscular receptors could be viewed as sensors of the state of visceral activity, that is, distention or contraction, whereas serosal receptors would act as sense organs responding to stretching or positioning of the entire viscera.

4.2.1.2 Rapidly Adapting. Mechanoreceptors that adapt rapidly to maintained mechanical events are principally located in the superficial layers of the viscera and in association with their serous covering and supports. They connect centrally by both myelinated and unmyelinated fibers and include the largest diameter visceral fibers, the Aβ fibers of pacinian corpuscles.

Rapidly adapting mechanoreceptors do not respond to the forms of stimulation that typically activate the slowly adapting type, such as distention, contraction, or stretching (Fig. 9). Therefore, they are not active under a variety of physiological circumstances but respond to more specific situations, such as those associated with certain postural rearrangements or abnormally vigorous activity of the viscus. In addition to the serosal type of receptor, a certain number of mucosal rapidly adapting mechanoreceptors have been described; these respond to well-defined stimuli, such as stroking of the mucosa of the digestive tract (Leek 1969; Harding and Leek 1972a, b; Clerc and Mei 1983).

Similar receptors located in the mucosa have also been demonstrated in the respiratory tract (Widdicombe 1954a, b; Fillenz and Widdicombe 1972). As a rule, surveys reveal that rapidly adapting mechanoreceptors are encountered less often than the slowly adapting varieties. This could be related to difficulties in determining the appropriate stimulus and may not indicate a true difference in relative numbers.

Apparently the rapid adaptation noted for mechanical stimulation is not a feature of the nerve fiber as such, for when mucosal rapidly adapting receptors

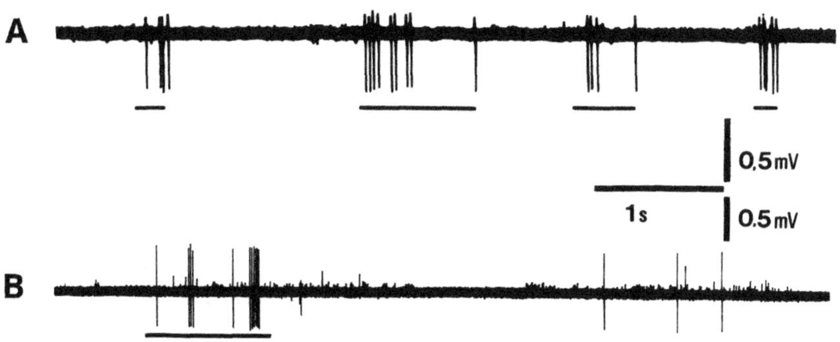

Fig. 9A, B. Digestive mucosal mechanoreceptors. **A** Receptor located in the duodenum stimulated by mucosal stroking; **B** receptor located in the lower esophageal sphincter activated by digital compression. Note the presence of activation after the stimulation (Clerc, unpublished). In both cases, stimulations are indicated by lines

are chemically excited they show a relatively slowly adapting discharge (Clarke and Davison 1978).

4.2.2 Chemoreceptors

Until fairly recently, the only well-established visceral chemoreceptors were the arterial type, found in the carotid body or in the aortic region, which are excited by a decrease in P_{O_2} and in increase in P_{CO_2}. However, chemoreceptors are now known to be present in the digestive tract the liver, and the kidney. Several types of chemoreceptors have been described for the digestive tract, most of which have unmyelinated fibers.

4.2.2.1 Alkali- and Acid-Sensitive Receptors. A class of alkali- and acid-sensitive receptors have fibers in the vagus nerve (Iggo 1957b). These elements are not strictly pH detectors, since, for a number, the nature of the acid or alkali providing the stimulus influences the response (Leek 1977; Clarke and Davison 1978; El Ouazzani and Mei 1981).

4.2.2.2 Glucoreceptors. Glucoreceptors are located in the stomach (El Ouazzani and Mei 1981) and a larger number in the small intestine (Sharma and Nasset 1962; Hardcastle et al. 1978; Mei 1978; Perrin et al. 1981). Those located in the gastric and duodenal region have afferent fibers in the vagus (Mei 1978; El Ouazzani and Mei 1981). Those situated in the jejunal region and a certain number of duodenal units have afferent fibers passing through the splanchnic nerves (Sharma and Nasset 1962; Hardcastle et al. 1978; Perrin et al. 1981). Such afferent units appear to be most responsive to glucose solutions, although they can also be activated by other carbohydrates. Each unit possesses a specific spectrum of sensitivity for the different carbohydrates (Hardcastle et al. 1978; Mei 1978).

4.2.2.3 Amino Acid Receptors. Amino acid receptors with afferent fibers in the vagus are known to exist in the small intestine of cats (Jeanningros and Mei 1980; Jeanningros 1982). In the same manner as the glucoreceptors, the amino acid-sensitive units are generally quite specific, in that they are not activated by other chemical, osmotic, thermal, or mechanical stimuli (Fig. 10).

All of the chemoreceptors in the digestive tract show either an absolute or a relative specificity in their responses to different nutrient groups or agents. They also share a lack of responsiveness to mechanical, osmotic, or thermal stimuli, as is illustrated by the unit in Fig. 11. In spite of this apparent specificity, the response can be altered by factors that modify the absorption of nutrients or their accumulation (absolute luminal concentration, difference in concentration between the intestinal lumen and the bloodstream, intestinal motility, and intestinal temperature, for example (Fig. 12). The exact locus of the receptive terminals for intestinal chemoreceptors has not been determined; however, in view of their short latency of response to changes in mucosal concentration of substances, they are likely to be located in the neighborhood of the mucous membrane. They are probably associated with the many undifferentiated nerve bundles that have been

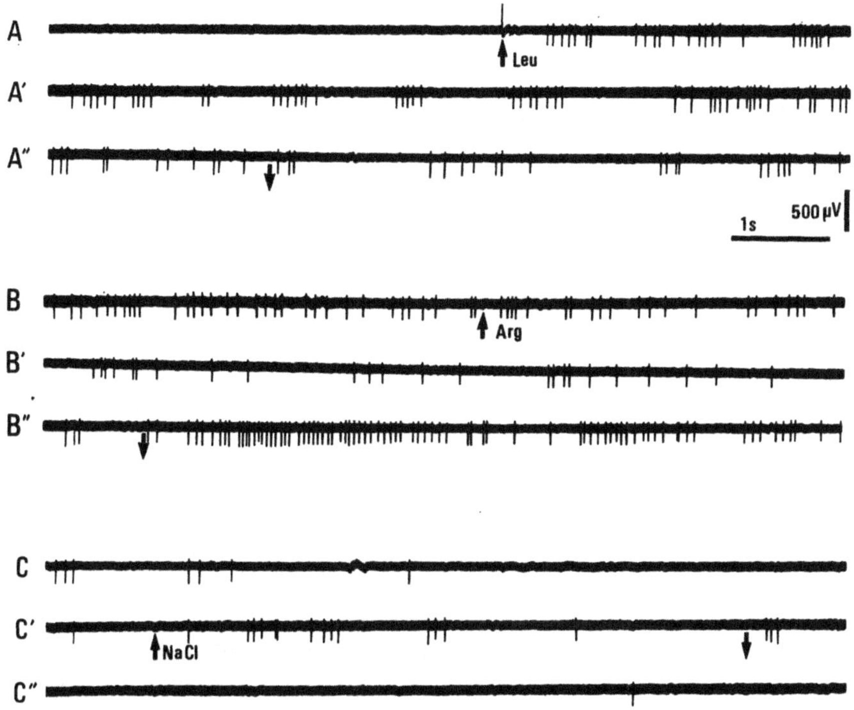

Fig. 10 A – C. Vagal amino acid receptor located in the duodenum of the cat. Effect of two amino acids (leucine and arginine, 250 mosmol/l) and osmotic pressure. *Arrows* indicate the beginning and the end of intestinal infusion. **A**, **A′**, and **A″**; **B**, **B′**, and **B″**; and **C**, **C′**, and **C″** are each continuous. **A**, **A′**, **A″**, response produced by leucine; **B**, **B′**, **B″**, response produced by arginine. The basal activity is due to previous stimulation; here the response occurs later. **C**, **C′**, **C″**, infusion with NaCl having the same osmotic pressure as the amino acids. No obvious response is visible (Jeanningros 1982)

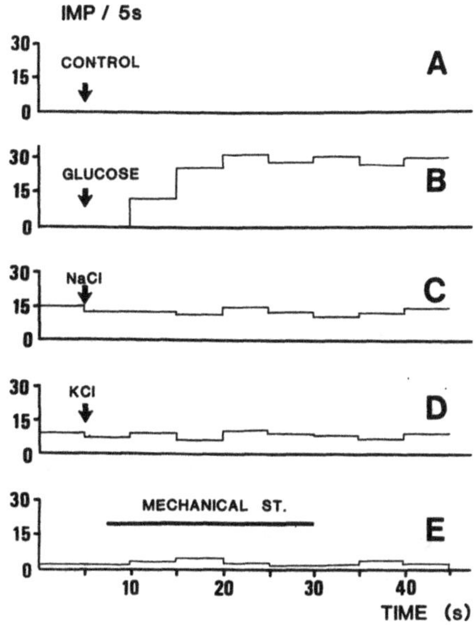

Fig. 11 A – E. Histogram illustrating the specificity of an intestinal glucoreceptor towards chemical, osmotic, and mechanical stimulations. **A** Control; **B** glucose solution (550 mosmol/l); **C** NaCl (550 mosmol/l); **D** KCl (550 mosmol/l); **E** strong distension of the whole duodenum; the line indicates the duration. In **B**, **C**, and **D**, 20 ml of solution were infused with a syringe during 10 sec; *arrows* show the end of infusion

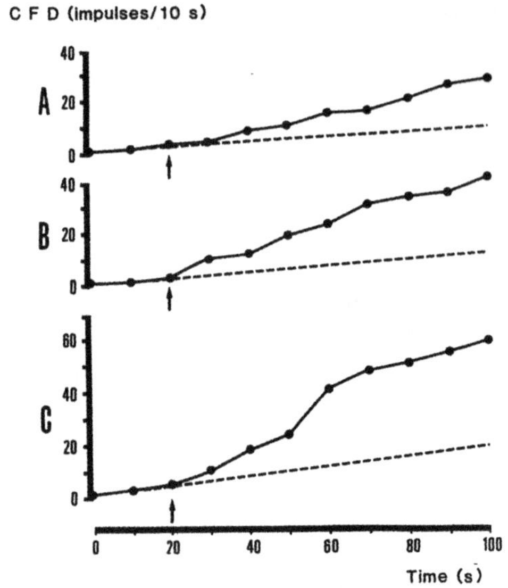

Fig. 12A–C. Effect of the glucose concentration on a splanchnic glucoreceptor. **A** 10 g/l; **B** 50 g/l; **C** 100 g/l. The cumulative frequency of discharges (*CFD*) increased from **A** to **C**. *Arrows* indicate the beginning of perfusions (Perrin et al. 1981)

described in the mucosa and the submucosa (Mei 1978). Nevertheless, special cells (recently discovered in the ileal mucosa by Newson et al. 1982) would represent a type of chemoreceptor. In general, intestinal chemoreceptors do not show ongoing activity in the absence of their specific stimulating agents. The response adapts slowly to maintained concentrations of the effective chemical agent, and the discharge rate rarely exceeds 20 impulses/s.

4.2.2.4 Other Chemoreceptors. Liver glucoreceptors, discovered by Niijima (1969a), differ from the intestinal type. They have an ongoing discharge that decreases when the concentration of glucose in the hepatic circulation increases. Chemoreceptors of the kidney, with characteristics quite different from those of the intestinal tract, have been described in anesthetized rats (Recordati et al. 1978, 1980). They respond to renal ischemia and urine composition but are not excited by changes in intrarenal pressure.

4.2.3 Thermoreceptors

Visceral thermoreceptors were first suspected when evidence was noted of thermoregulatory reactions initiated from the abdominal cavity (Rawson et al. 1969; Rawson and Quick 1972). Subsequently, unitary discharges associated with warmth receptors that were active at normal temperatures were discovered in mesenteric nerves (Riedel 1976). More recently, Gupta et al. (1979) described cold receptors in splanchnic nerves in the cat gastrointestinal tract. Warmth, cold, and a mixed variety of thermoreceptors seem to be widespread in the digestive system of the cat; they have been noted in the esophagus, stomach, and the small intestine (El Ouazzani and Mei 1979, 1982). An example of their activity is shown in Fig. 13. In the cat, the fibers of these various thermoreceptors are

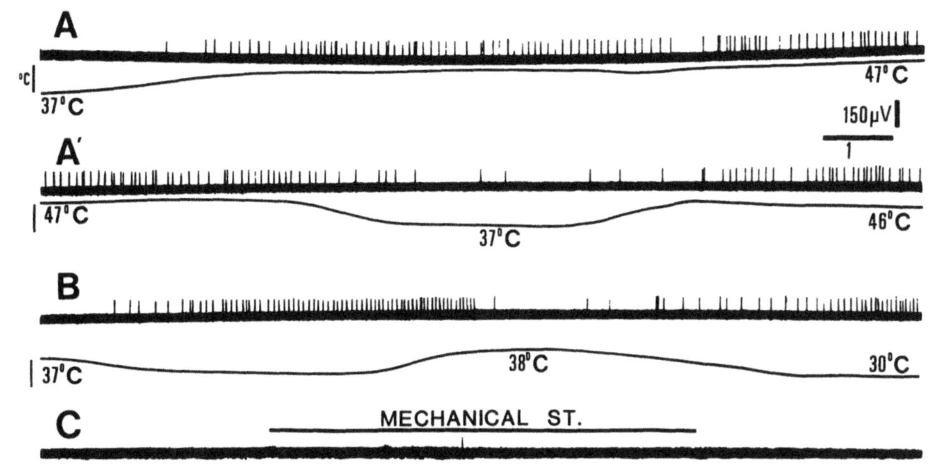

Fig. 13 A – C. Mixed thermoreceptor located in the antroduodenal area of cat. **A** The receptor, which is silent at 37 °C, discharges during heating of the antroduodenal area; **A′** when the temperature returns to the normal value (37 °C), activation disappears almost completely. After a second heating (*right*), the receptor is again activated; **B** the receptor responds to 2 cold stimulations (30 °C); **C** compression of the antroduodenal area does not induce any response (El Ouazzani 1981)

found in the vagal nerve, and the receptors are probably localized in the superficial layers of the digestive tube. The vagal thermoreceptive fibers have characteristics resembling those of the cutaneous and lingual distributions, in that they adapt slowly and develop similar maximal discharge frequencies relative to temperature in an inverted bell-shaped curve. However, unlike the thermoreceptive fibers in the epidermis, they do not show a basal activity at normal temperatures for the tissue. The vagal thermoreceptors are apparently almost always associated with unmyelinated afferent fibers.

4.2.4 Osmoreceptors

The existence of receptors responsive to osmotic pressure changes had been postulated on the basis of several studies dealing with gastric emptying (Hunt and Knox 1968). However, such receptors have not yet been electrophysiologically confirmed. Moreover, it is not even certain that the information on alterations of osmotic pressure derives exclusively, or even primarily, from specific osmoreceptors, since some mucous membrane mechanoreceptors are activated by hypertonic solutions (Leek 1977). In fact, in experimental work being conducted in my laboratory, we have noted that certain mucosal receptors of the small intestine were excited by either hypo- or hypertonic solutions in the lumen. Most of these receptors should not be classified as true osmoreceptors for two reasons: (a) their responses are not purely a function of osmotic activity but depend also on the nature of the osmotically active substance, i.e., glucose, mannose, sorbitol, NaCl, etc.; and (b) they usually respond to several forms of mechanical, as well as chemical and thermal, stimuli (Garnier and Mei 1982). Therefore, these ele-

ments would appear to have a multimodal responsiveness (see Sect. 4.2.5). However, other receptors which are not sensitive to mechanical and thermal stimuli might represent a kind of true osmoreceptor.
Apparently, specific osmoreceptors do exist in the liver; they are the only visceral units that have been documented with characteristics appropriate for this classification (Niijima 1969b; Andrews and Orbach 1975; Adachi et al. 1976). The hepatic osmoreceptors have a discharge whose frequency is proportional to the osmotic pressure of the solutions employed, regardless of the nature of the osmotically active agent (NaCl, sucrose, etc.).

4.2.5 Multimodal Receptors and the Specificity of Enteroreceptors

A group of afferent units with receptive terminals in the gastrointestinal mucous membrane apparently responds indiscriminately to different classes of stimulation, such as mechanical, thermal, chemical, or osmotic (Harding and Leek 1972a, b, 1977; Clarke and Davison 1978). Present evidence suggests that in the somatic region, such nonselective receptors exist only as nociceptors. "Nociceptors are distinguished by elevated thresholds to all kinds of stimuli and at least one cutaneous type." Polymodal nociceptors with unmyelinated fibers respond indiscriminately to any stimulus that is sufficiently intense to threaten the physical integrity of the tissue. It is not known whether the multimodal type of intestinal receptor is analogous to the polymodal cutaneous nociceptor, since defining noxious stimuli for visceral structures is not a simple task. The very exposure of a viscus for experimentation or recording of afferent discharges is probably a noxious event, since these structures are normally protected from dry air and vigorous mechanical manipulation but when exposed are inevitably subjected to these stimuli. Further, if one accepts the proposition that nociceptors are the sense organs leading to painlike reactions, it is noteworthy that the intestinal tract itself generally does not give rise to pain; pain of intestinal origin is usually attributed to afferent activity from mesenteric structures.
As a result, it is tempting to propose that the visceral territory is provided with two general types of afferent units: (a) specific afferent units, such as arterial or digestive chemoreceptors or low threshold mechanoreceptors, and (b) nonspecific afferent units, such as the multimodal intestinal receptors. A similar situation apparently exists in the unmyelinated sensory units of the bronchopulmonary area, although the normal or adequate stimuli in this region have still not been established (Coleridge et al. 1978; Delpierre et al. 1981).
Admittedly, it is difficult at times to distinguish between the two types, particularly since there is more than one kind of specific receptor, and the units differ in the degree of their specificity.
The meaning of the activity of a single neuron that responds to more than one kind of stimulus should be considered. For example, the cardiopulmonary receptors, which are located in the vicinity of a large, pulsating blood vessel or the heart itself, may be active in relation to both pulmonary inflation and changes in cardiac output. Similarly, the multimodal gastrointestinal receptors respond to both mechanical events in the intestine and certain chemical stimuli. Is the receptor's activity generated by two different classes of equally important stimuli, or

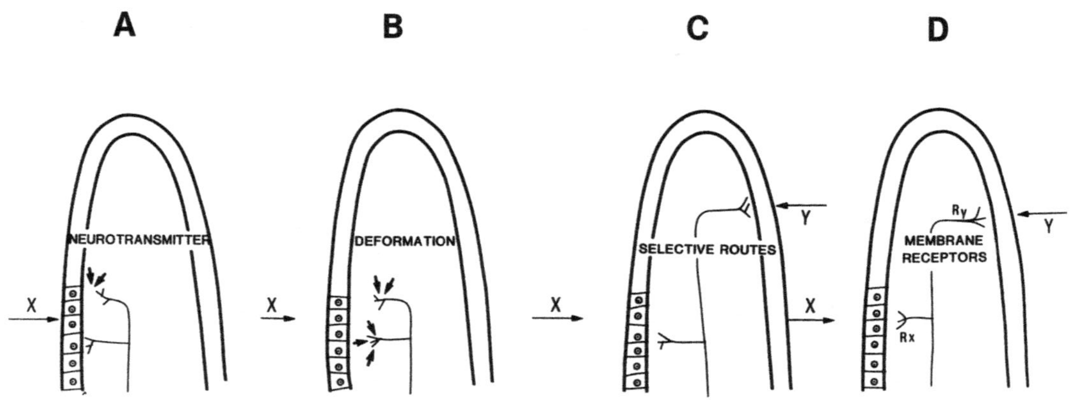

Fig. 14A–D. Putative mechanisms involved in chemoreceptor activation. **A** Neurotransmitter (or neuromodulator) release by the stimulating substance (*X*); **B** deformation produced by the stimulating substance (*X*); **C** each stimulating substance (*X* and *Y*) follows selective routes; **D** each terminal nervous branch possesses membrane receptors (*RX* for substance *X* and *RY* for substance *Y*). In the diagram, intestinal villi and mucosal receptors have been represented, but mechanisms **A** and **B** were first proposed for arterial chemoreceptors

are there primary and secondary types of stimuli? If primary and secondary manners of activating a sensory neuron exist, does the secondary stimulus generate distracting activity (i.e., noise) in the central nervous system? At present there is no answer to these questions; however, it is clear that an understanding of the matter will depend upon knowing how different sense organs respond to equivalent stimuli, as well as how the CNS manages such information.

One thing appears likely: the gross structure of afferent receptors in viscera cannot account for their functional specificity. With the exception of the carotid and aortic chemoreceptors and the mesenteric pacinian corpuscles of the cat, complex and specialized structural arrangements are not ordinarily found. How can specificity be explained in the absence of evident structural differentiation? Two possibilities have been proposed in the literature:

1. A depolarization and consequent generation of impulses may be produced by a neurotransmitter. This hypothesis presumes that a cell or structure is presynaptic to the afferent nerve terminal. Such an arrangement is compatible with the structure of chemoreceptors, such as those of the carotid body, that have associated with their afferent nerve terminals well-defined nonneural cells. One type of these associated cells has specialized metabolic characteristics. It should be noted that definite involvement of the metabolic cells in the afferent function of the carotid or aortic bodies has not yet been established (Biscoe 1971; Hayashida et al. 1980). In one case, this postulated arrangement is not appropriate for the digestive tract chemoreceptors, which apparently consist of undifferentiated nerve endings. In the latter situation, the concept of a mediating substance can be reconciled by presuming diffuse release into tissue or local circulation by specialized structures located elsewhere (Fig. 14A).
2. The chemical stimulus might cause a mechanical deformation, a concept proposed by Paintal (1968) for arterial baroreceptors. Chemoreceptors in this scheme would simply represent a form of mechanoreceptor (Fig. 14B). This interesting possibility is yet to be examined experimentally.

As alternatives to these hypotheses, we propose two theories to explain chemoreceptor specificity which take into account the absence of associate cells for certain sensory units. In both concepts, the chemical agents are presumed to act directly on the afferent nerve terminals.

In the first theoretical case, the receptive terminals of intestinal chemoreceptors are specially placed relative to the absorption pathway for nutrients (Fig. 14C). This theory is based upon ultrastructural studies which show the villi of the intestine to contain nerve endings, without surrounding specializations, that terminate at different loci and at different levels relative to the lumen. In the second theoretical case, the membrane of chemosensitive endings, like that of the gustatory and olfactory receptors, may be equipped with membranous acceptors capable of selectively fixing a given substance (Fig. 14D). In the present state of the art, our two theories seem to account best for the functional characteristics of intestinal chemoreceptors, especially their specificity.

5 Functional Significance of Visceral Afferent Messages

For a long time, afferent activity from the viscera was not considered as important as that stemming from somatic structures. Factors contributing to this idea included the substantial underestimation of the number of visceral afferent fibers, the limited number of distinct sensory experiences related to visceral structures, and the dominance of pain as a sensory experience derived from the viscera. Until some two decades ago, the prevailing idea was that parasympathetic afferent input was principally concerned with regulation of physiological activity related in one way or another to the viscera, and that afferent activity passing through the sympathetic nerves was associated with situations of stress and pain. Our present understanding is different and is based upon a number of studies that have uncovered an amazing complexity in mechanisms which depend, in part, upon enteroreceptor activity. Because these are recent discoveries, significant gaps remain in our understanding of the relationship between electrophysiological observations on the activity of enteroreceptors and the functional arrangements into which they enter. The part played by several established types of visceral receptors has yet to be established. Similarly, a number of known reactions have not been tied to particular types of afferent units and pathways. One of the problems is that experiments based upon transection and electrical stimulation of visceral nerves often produce variable or confusing results; such procedures are useful mainly in defining the efferent pathway for a reflex. For other purposes, when responses are evoked by electrical stimulation of nerves, it is difficult to dissociate the roles of neural and humoral mediation.

It is possible to divide reactions of visceral origin into two main groups, those associated with physiological regulation and those appearing under pathological circumstances. Another classification distinguishes between visceral reactions associated with conscious awareness and those that operate entirely at subconscious level. As a general rule, subconscious reactions are part of normal physiological processes, and those that reach conscious recognition appear to be associated mainly with feelings of discomfort or pain. For this reason, we shall divide visceral reactions into those associated with physiological regulation and those associated with protective or nociceptive reflexes. Our special attention will be given to physiological mechanisms associated with the digestive tract, since again recent advances in understanding have been related to this system.

5.1 Physiological Regulatory Mechanisms

Physiological reactions and regulations will be examined in terms of how they trigger and coordinate visceral activity, provide for homeostasis, and coordinate skeletal-visceral activity (behaviorally related mechanisms).

5.1.1 Triggering and Coordinating Visceral Reactions

Mechanisms initiating a visceral reflex or controlling the interaction between visceral reflexes are the best known, since they can be readily induced under experimental conditions and involve relatively restricted neurocircuitry. Further, they can be characterized according to the distribution of the efferent activity, that is, whether or not the efferent activity involves only the effector organ from where the afferent signals originate, or whether other effector structures are directly influenced. Therefore, one can refer to these as either limited (or single organ) reflexes or extensive (or multiorgan) reflexes. It is noteworthy that afferent signals carried by myelinated fibers play a large part in such triggering and coordinating reactions.

5.1.1.1 Reflexes of Limited Distribution. Certain visceral organs or systems of organs, such as those involved in respiration and circulation of blood, exhibit continual activity, whereas others, such as those of the digestive or genital systems, have periodic or episodic activity. Reflexes that affect a limited group of effector organs are part of the functional arrangement for all visceral organs, both the continually active and the periodically active. A well-known and fairly typical example of this kind or reflex is the one originating by the vagal pulmonary stretch receptors, a reflex which modulates the frequency and depth of breathing through an action upon the medullary respiratory centers. Another limited reflex begins with enhanced activity of the digestive system mechanoreceptors which then evoke a tonic increase in the basal level of intestinal smooth muscle tone (background contraction), as well as particular phasic effects; e.g., enhanced activity of the esophageal mechanoreceptors in the esophageal muscular coat contributes to deglutition by facilitating the centrally organized progression of excitation down the esophagus (Roman and Gonella 1981). The latter example is diagrammed in Fig. 15A.

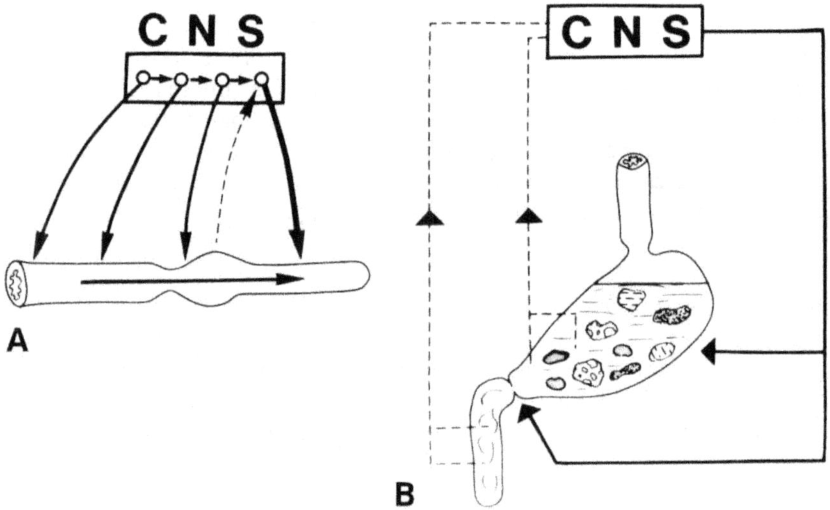

Fig. 15A, B. Role of digestive receptors in triggering and coordination of digestive motility. **A** During the passage or peristaltic wave (*arrow*), esophageal mechanoreceptors provide information which may reinforce the central output. **B** Gastric emptying partly depends on various nervous signals originating from both stomach and duodenum

5.1.1.2 Distributive or Multiorgan Reflexes. Multiorgan reflexes exist in the situation in which several different parts of a complex system, such as the digestive tract, participate in an activity that is sequential or interrelated. For instance, the afferent input from the mechanoreceptors of the esophagus elicits relaxation of the lower esophageal sphincter. The same food intake initiating this esophageal reflex evokes an enteroenteric reflex and an enterocolic reflex, both excitatory, and an enterogastric inhibitory reflex (Gonella 1978; Roman and Gonella 1981). These multiple reactions involving different effector portions of the digestive system are good examples of the complexity of the extended reflexes and the participation of enteroreceptors excited by different stimuli. As Fig. 15B illustrates, a single process such as gastric emptying is the result of afferent input from mechanical, osmotic, and probably thermal stimuli; it produces a mechanical and chemical separation of the gastric contents. This concept of gastric emptying has evolved from observations that mechanical, chemical, osmotic, and thermal stimulation each produce a decrease in the motor activity of the stomach (Hunt and Knox 1968; Roze et al. 1977; El Ouazzani and Mei 1979; Roman and Gonella 1981). The importance of the afferent inflow is suggested by the fact that in vagotomized animals and human beings gastric emptying unequivocally slows (Brooks 1977). This slowing of gastric emptying in the absence of vagal afferent pathways seems likely due to a lack of input from the vagal mechanoreceptors, chemoreceptors, and thermoreceptors located in the small intestine, although dissociation of afferent and efferent effects in this case may be difficult.

The effects initiated by the carotid and aortic arterial chemoreceptors illustrate an example of a distributed visceral reflex interrelating different visceral systems. This set of reflexes produces changes in the respiratory control system, modulating ventilation through skeletal muscle effectors (diaphragm and intercostal muscles), and also in the smooth muscle resistance of the arterial tree. The increased chemoreceptor activity from P_{CO_2} increase and P_{O_2} decrease in the systemic arterial supply initiates enhanced frequency and depth of respiration, as well as increased arterial tone and increased mean arterial pressure.

In considering the various widely distributed reflexes, it must be kept in mind that the various different classes of afferent elements of a particular region and structure do not necessarily elicit similar or cooperating reflexes. In some instances, afferent fibers arising from sense organs of one locus, but passing centrally via different pathways (i.e., parasympathetic or sympathetic nerves), evoke opposite effects. For example, increased activity of the afferent fibers of digestive tract receptors in the vagus nerve augments respiration (Mei 1976), whereas increased activity of fibers passing centrally in the splanchnic nerves from the same general region inhibits respiratory activity (Albano and Garnier 1977). Furthermore, although our examples have so far concentrated on reflexes modulating motor activity or the viscera, secretory activity is also under visceral reflex control. Cases in point are (a) an increase in secretion of gastric juice initiated by stimulation of mechanoreceptors of the stomach (Stenquist 1979) and (b) the modulation of pancreatic exocrine secretion by the concentration of amino acids in the intestine (Singer et al. 1980). The latter reflex might begin with excitation of mucosal chemoreceptors of the intestine that transmit signals in the vagus nerve and reflected evokes reflex excitation of parasympathetic efferent fibers distributed in the pancreas.

5.1.2 Homeostasis

Our understanding of the afferent limb of mechanisms controlling the constancy of the bodily environment has been considerably advanced recently as a result of increased knowledge about the afferent signals conveyed by unmyelinated fibers. The various regulatory processes involved in homeostasis have in common complex efferent pathways, regardless of the specific purposes they serve. The efferent control of these complex regulatory processes can be neural, humoral, neural-humoral, and ultimately may involve all three methods of modulating effector activity. Regulatory mechanisms of homeostasis often imply multiple levels of the central nervous system: spinal, medullary, mesencephalic, and hypothalamic. They can be characterized according to the nature of the process they regulate. For our purposes, four categories will be defined and considered: mechanisms controlling (a) arterial pressure, (b) endocrine secretion, (c) water and salt balance, and (d) body temperature.

5.1.2.1 Regulation of Arterial Pressure. Receptors sensing arterial pressure, the baroreceptors of the carotid sinus, the arch of the aorta, and several other large arteries, are known to play a vital part in regulating systemic arterial pressure. Recently, it has emerged that sensory receptors located in the kidneys, the liver, and the intestines may be concerned with regulation of the circulation to the organ in which they are situated. However, with the exception of the renal and hepatic receptors, identification of these newly uncovered afferent sources has not as yet been made. It is yet to be learned whether the sense organs in question operate as true pressure sensors or detect other features. For example, there is reason to believe that certain gastrointestinal mechanoreceptors activated during the process of digestion participate in reflex arrangements that cause a redistribution of blood between the intestine and other vascular beds.

As mentioned above, some general differences exist in the reflex effects produced by afferent fibers that course through parasympathetic and through sympathetic nerves. Increased signals in afferent fibers of parasympathetic nerves tend to produce hypotensive effects (Paintal 1955), whereas increased signals in afferent fibers passing through sympathetic nerves tend to have a dominant hypertensive result (Reiman and Weaver 1980). However, it is clear that exceptions do exist, such as the hypertensive effect of the carotid and aortic chemoreceptors, whose afferent pathways parallel those of the baroreceptors arising from the same regions.

5.1.2.2 Endocrine Secretion. Efferent effects on endocrine secretion produced by stimulation of autonomic nerves have been known for some time. Labarre (1927) found that blood glucose levels decreased due to an increase in insulin release upon electrical stimulation of the peripheral end of the vagus nerve. As a consequence of the development of radioimmunoassay, such early observations have been confirmed, and general features of the effects have been established. Vagal efferent activity stimulates the release of insulin, as well as other endocrine secretions, by the pancreas (Woods and Porte 1974). Vagal efferent impulses also cause release of gastric (Stenquist 1979) and intestinal hormones (Bloom and Edwards 1980). In contrast, splanchnic efferent activity either has no effect upon

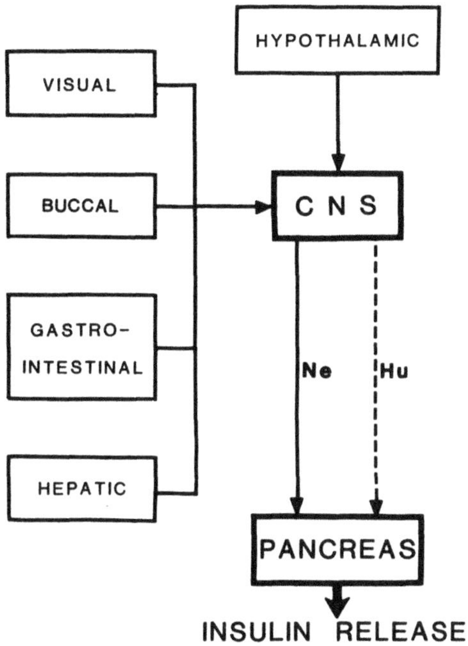

Fig. 16. Nervous pathways acting in insulin secretion, starting from visual, buccal, gastrointestinal, hepatic, and hypothalamic areas. *Ne*, efferent neural pathway; *Hu*, humoral pathway. Concerning the gastrointestinal glucoreceptors, pathways involve only nervous circuits (mainly vagus nerves, see Mei et al. 1981)

the endocrine output of these three regions or causes an inhibition of their endocrine activity. In some instances, the splanchnic efferent activity causing an inhibition of endocrine secretion apparently does not act directly upon the secretory cells but rather through a modulation of the blood supply to the structure (Bloom and Edwards 1980).

The existence of efferent effects does not necessarily indicate a control process initiated by enteroreceptors, although this has been established in the case of insulin. Excitation of glucoreceptors of the intestine with afferent fibers in the vagus leads to a rapid increase of blood insulin levels (Mei et al. 1981) (Fig. 16). As far as we know, the principal function of the endocrine regulation mediated through afferent and efferent links in the autonomic nerves is to prepare or lay the ground for other control mechanisms and to potentiate the effects of other regulatory actions. In the case of the reflex action of the glucoreceptors upon insulin, the reaction seems to be preparatory for the direct action of glucose in causing insulin release. A more detailed discussion of anticipatory regulation can be found in a recent paper (Louis-Sylvestre 1978).

5.1.2.3 Water and Salt Balance. Classically, diuresis is thought to be produced by a rise in the central venous pressure and osmotic pressure or an increase in right atrial volume. The posterior pituitary is the principal effector pathway for this reflex by decreasing its release of antidiuretic hormone and indirectly, through its effect on renin production, by decreasing aldosterone activity.

The recent literature provides evidence that other peripheral parts of the autonomic system may be involved in this regulation. One of them includes hepatic vessels, since an osmotic pressure increase in hepatic circulation induces diuresis (see, in particular, Haberich 1968; Bie 1980; Rogers and Novin 1980;

Baertschi and Vallet 1981). It is conceivable that the hepatic osmoreceptors of the type described by Niijima (1969b), Andrews and Orbach (1975), and Adachi et al. (1976) represent the afferent limb of this diuretic reaction.
Still another modulation of urine production might be initiated by the introduction of hypotonic or hypertonic solutions into the gastrointestinal tract (Kill and Andersen 1972; Bie 1980). In this case, the intestinal osmosensitive receptors may represent the origin of the afferent limb.
It has recently come to light that afferent neurons of the heart, other than the atrial receptors mentioned above, may also participate in salt and water regulatory mechanisms (Reiman and Weaver 1980), but a more complete story has emerged for renal receptors that, like these cardiac ones, have afferent fibers in sympathetic nerves (Niijima 1971). From the present evidence, it appears that the renal afferent units participate in a localized and highly organized regulatory reaction, because the diuresis starting from renal receptors may control urine formation and salt secretion of the two kidneys independently. Denervation of one kidney with loss of its afferent input causes an increased diuresis in that kidney and a decreased diuresis in the contralateral kidney (Colindres et al. 1980). This latter observation provides reason to believe that the renal afferent pathways have a tonic effect upon the renal function. Renorenal reflexes involve not only a neural pathway but also an efferent link through a hormonal mechanism, since action upon the renal tubules apparently takes place through the medium of aldosterone.

5.1.2.4 Thermal Regulation of Visceral Origin. The recent direct evidence on the existence of specific thermoreceptors in the visceral region (see Sect. 4.2.3) is consistent with and supports observations by Riedel (1976) in the rabbit and Rawson and Quick (1972) in the sheep that describe thermoregulatory reactions following localized heating of the abdominal cavity. It is also consonant with the observation that artificial abdominal cavity warming increases food intake in the rat (Di Bella et al. 1981). The functional significance of thermal regulation from the viscera is not entirely clear. It seems likely that such mechanisms have a role in overall thermal regulation whenever the temperature in the digestive tract rises above or falls below the optimum temperature, as would occur during the ingestion of food. Thermoregulatory responses from the abdominal cavity are more readily understood in ruminants, since the fermentation that is normal for the gastric contents raises the temperature there by several degrees. It is significant that thermoreceptors of the vagus nerve also have a role in the coordination of digestive motility (El Ouazzani and Mei 1979). Finally, the notion that thermoreceptors of the gastrointestinal tract and liver can influence food and fluid intake in all species is in agreement with the observations mentioned above for the rat (Brobeck 1960; Di Bella et al. 1981).

5.1.3 Visceral-Somatic Interactions

It has long been evident that reactions mediated by the autonomic nervous system are involved in the coordination of integrated motor activity that makes up behaviorally manifest responses; such complex manifestations of activity depend upon coordination of respiratory, cardiovascular, excretory, and digestive functions with skeletal motor activity. Study of physiological changes and behavioral manifestations have provided abundant evidence on the diversity and intricacy of the integrated mechanisms involved (Sato and Schmidt 1973). One of the simpler examples is that a decerebrate animal can be made to adopt the postures associated with defecation or copulation by stimulating afferent structures respectively of the rectum or the genitalia. Another case is the fact that bladder distention in spinal animals and human beings will, under certain conditions, evoke contractions in the muscles of the abdomen and the hindlimbs consistent with activity associated with voiding. In a situation relatable to nociception and protective reactions, skin pinching has been established to inhibit gastrointestinal motility (Sato and Terui 1976; Kametani et al. 1979).

On the other hand, in a more global sense the functional significance of visceral afferent input under normal conditions remains largely speculative. There is solid evidence that visceral afferent input influences sleep and wakefulness. Vagal afferent input (Bonvallet et al. 1954) and activity from the aortic region (Puizillout and Foutz 1977) and the stomach (Garnier 1968) tend to be soporific. In contrast, stimulation of afferent fibers of the splanchnic nerve produces signs of waking in the electroencephalogram from the cerebral cortex; whether or not this alerting or waking effect is solely related to the nociceptive fibers of the splanchnic nerves or can also be produced by non-nociceptive afferent input has not been established.

The present discussion has emphasized afferent organization from the alimentary tract, for which a fairly detailed understanding of behavioral effects is emerging. At one time, based upon observations on vagotomized animals and human beings, it was proposed that afferent input from the digestive tract was not essential for digestion and alimentary function. This proved to be too sweeping a conclusion, as happened in many other situations analyzed on the basis of effects produced by lesions of the nervous system. At present, it is known that afferent input from the gastrointestinal tract, principally vagal, is important for short-term control of food intake (Powley et al. 1976; Fox et al. 1976; Novin 1976). Thus, the intestinal chemoreceptors appear to be the initiating point for satiety induced by sugars or proteins prior to absorption of all of the digestive products (Nicolaidis 1978). Glucoreceptors of the intestine with vagal afferent fibers apparently participate in the pleasurable sensations (alliesthesia) related to sugar intake (Cabanac and Duclaux 1973; Cabanac and Fantino 1977). Comparison of the food intake after intestinal perfusion with one of the many forms of nutrient substances (carbohydrates, protein, fats) in normal and chronically vagotomized rats provides evidence that afferent activity related to glucose and amino acid concentrations does play a part in the control of food intake (Novin et al. 1979).

It appears reasonable to believe that an increased understanding of the visceral afferent contribution to various mechanisms regulating complex behavior will help to provide explanations for dysfunction and metabolic disorders found in the clinic. This will be particularly true for human beings with bilateral vagotomy who may exhibit serious disturbances such as postprandial dumping (Rignault 1971), loss of body weight and underconsumption (Fox et al. 1976), or decrease in glucose tolerance (Woods and Porte 1974).

5.2 Protective and Nociceptive Reflexes of Visceral Origin

5.2.1 Protective Reactions

Since protective reflexes appear only under specific, instigating circumstances, one may conclude that they are initiated by sense organs with no background or "spontaneous" activity. Several sense organs that are silent under normal physiological circumstances appear to have myelinated afferent fibers. This is the case for vagal nerve afferent units in the respiratory and digestive tracts. It is possible to argue that, from a teleological standpoint, protective reactions need to occur rapidly, and therefore the afferent signals must be conducted in larger diameter fibers. Nevertheless other endings involved in protective reactions are also connected to C. fibers.

Vagal mechanoreceptors with relatively elevated thresholds that are located in the gastrointestinal tract (serous membrane) are the receptors most likely to be activated in vomiting. They and certain other mesenteric receptors can evoke apnea (Albano and Garnier 1977). Apnea would reduce the possibility of inhalation of vomitus. Respiratory tract receptors excited by irritant substances evoke coughing, which causes expulsion of the irritant substances such as dust, smoke, and mucus from the respiratory tree.

During moderate to violent exercise, a reflex causing respiratory activation and a reduction in somatic muscle activity is initiated from pulmonary tract receptors of the J type (Anand and Paintal 1980). These last two examples of protective reflexes are instances of generalized effects that extend to several effector organs and alter the behavior of the entire organism.

5.2.2 Nociceptive Reactions

On the basis of clinical and early experimental observations, it was long believed that afferent fibers in sympathetic nerves were those that were largely responsible for pain and the reflexes associated with it. This idea is only partially correct. It is true that in the digestive system afferent activity in the splanchnic nerves is more likely to produce painlike reactions than is that in the vagus nerves; however, certain forms of gastrointestinal discomfort associated with ulcers apparently originate from activity in vagal afferent fibers. It is also true that the majority of pain and nociceptive reactions from the respiratory system and the pelvic region are initiated by afferent fibers passing through nerves that would be classified as part of the parasympathetic organization.

A perennial question arises about the initiation of pain in the case of visceral receptors. Does this begin with specific sense organs, that is, nociceptors, or do sense organs excited by physiological levels of stimulation, provide some special kind of signal when subjected to strong stimulation? It seems unquestionably to be the case that, for the visceral regions like the somatic regions, there are a series of specialized sense organs that must be recognized as nociceptive receptors. For example, the receptors of the mucosal or serosal regions adapt rapidly to maintained stimulation and have high thresholds. Nociceptors are characterized by elevated thresholds to all forms of physiological activity that would normally stimulate other sense organs of the region. On the other hand, "physiological" sense organs can also be involved in any sensory phenomenon or reflex, if they provide quantitatively different signals when noxious stimulation is used. Unfortunately, in most instances, careful consideration as to what is noxious and what is not has rarely been given in investigations of the effects of direct stimulation of visceral structures. The essential question is whether or not the signals produced by strong stimulation of noxious intensity differ in some consistent fashion from those produced by any form of nonnoxious stimulus. Only under the latter circumstances would there be clear differentiation as far as the central nervous system goes. It must also be recognized that intense stimulation will certainly produce an ensemble of input that is likely to differ from that initiated by any normal physiological events. Consequently, under circumstances of intense stimulation, the overall activity of low threshold sense organs could differ significantly from that found in innocuous circumstances. Thus, from the visceral region, as in other parts of the organism, nociceptive reactions and nociceptive pathways are probably responsible for the reactions that are classified as pain producing and pain associated. However, they are certainly modified by afferent activity from receptors responsive to physiological events.

6 Conclusions

The various studies of visceral afferent organization and function over the past several decades have served to emphasize its importance and diversity. The following summarizes several of the major points that have emerged.

Visceral afferent fibers are far more numerous than had been previously believed and in many visceral nerves, either parasympathetic or sympathetic, may outnumber the efferent fibers. The ratio of unmyelinated to myelinated afferent fibers is particularly high in visceral nerves. When one considers the difficulty in recognizing and counting unmyelinated fibers in light-microscopic studies, the earlier underestimation of the visceral afferent component can be readily understood. The concept once held by some that the visceral afferent system was relatively homogeneous has also been dispelled. Afferent fibers of all diameters are represented in visceral nerves and include myelinated fibers in the Aβ, Aγ, and Aδ (or B) range, as well as the unmyelinated C fibers. It is important to recognize that differences in diameter occur along primary afferent neurons in the visceral region; fibers that are unmyelinated over one part of their course may become myelinated for the remainder. Analysis of the morphology of enteroreceptors at the ultrastructural level is still in its early stages, but at the light level, it is clear that diverse complexities in the terminal structures are not common.

Electrophysiological evidence has established that there is great variety among the primary afferent units associated with visceral structures. Most of the afferent activity from the viscera is carried by unmyelinated fibers, and they provide substantially detailed information. Individual afferent units often display a considerable specificy in the nature of the events they signal. It has become clear that every physiological mechanical, thermal, or chemical event activates at least one type of enteroreceptor, so that the afferent information arriving from the visceral region is remarkably complex. The best information on this point has come from studies on the digestive tract, for which different kinds of receptors sensing (a) mechanical, (b) chemical, and (c) thermal changes have been discovered. The specificity noted for the enteroreceptors is quite variable; the most specialized and selective ones include the thermoreceptors and certain chemoreceptors, and the less specialized are those that have a distinct, multimodal responsiveness to several different kinds of mechanical and chemical changes. In spite of major difficulties in establishing proper conditions for analyses because of multiple feedback and effector arrangements, it has emerged that visceral afferent input is important for a variety of systemic, visceral regulations, as well as visceral coordination in behavioral reactions involving the skeletal muscular system. These regulatory and behavioral reactions involve mediation by neural as well as neural-humoral pathways. Although at one time, the principal emphasis was the part of visceral afferent activity involved in protective reflexes and pain, it now appears that these are less important than its role in physiological regulation. It seems certain that the functional organization of visceral afferent mechanisms will prove to be one of the most exciting and rewarding areas of future study.

References

Adachi A, Niijima A, Jacobs HL (1976) An hepatic osmoreceptor mechanism in the rat: electrophysiological and behavioral studies. Am J Physiol 231:1043–1049

Adrian ED (1933) Afferent impulses in the vagus nerve endings. J Physiol (Lond) 61:49–72

Agostoni E, Chinnock JE, Daly de Burgh M, Murray JG (1957) Functional and histological studies on the vagus nerve and its branches to the heart, lungs and abdominal viscera in the cat. J Physiol (Lond) 135:182–205

Akoev GN, Chelyshev YA, Elman SI (1976) Effect of acetylcholine and catecholamines on excitability of pacinian corpuscules. Progr Brain Res 43:187–193

Albano JP, Garnier L (1977) Effets respiratoires de la stimulation des afférences splanchniques chez le chat légèrement anesthésié. Arch Ital Biol 115:1–19

Anand A, Paintal AS (1980) Reflex effects following selective stimulation of J receptors in the cat. J Physiol (Lond) 299:553–572

Andrew BL (1954) Proprioception at the joint of the epiglottis of the rat. J Physiol (Lond) 126:507–523

Andrew BL (1957) Activity in afferent nerve fibres from the cervical oesophagus. J Physiol (Lond) 135:54–55

Andrews JH, Andrews WHH (1971) Receptors, activated by acid, in the duodenal wall of rabbits. Q J Exp Physiol 56:221–230

Andrews WHH, Orbach J (1975) Effects of osmotic pressure on spontaneous afferent discharges in the nerves of the perfused rabbit liver. Pfluegers Arch 361:89–94

Andrews PLR, Grundy D, Scratcherd T (1980) Vagal afferent discharge from mechanoreceptors in different regions of the ferret stomach. J Physiol (Lond) 298:513–524

Arlhac A (1972) Données nouvelles sur la décharge des mécanorécepteurs vésicaux. Pfluegers Arch 333:258–270

Aström A, Crafoord J (1968) Afferent and efferent activity in the renal nerves of cats. Acta Physiol Scand 74:69–78

Baertschi AJ, Vallet PG (1981) Osmosensitivity of the hepatic portal vein area and vasopressin release in rats. J Physiol (Lond) 315:217 – 230
Beacham WS, Kunze DL (1969) Renal receptors evoking a spinal vasomotor reflex. J Physiol (Lond) 201:73 – 85
Beaudreau DE, Jerge CR (1968) Somatotopic representation in the gasserian ganglion of tactile peripheral fields in the cat. Arch Oral Biol 13:247 – 256
Bessou P, Perl ER (1966) A movement receptor of the small intestine. J Physiol (Lond) 182:404 – 426
Bessou P, Burgess PR, Perl ER, Taylor CB (1971) Dynamic properties of mechanoreceptors with unmyelinated (C) fibers. J Neurophysiol 34:116 – 131
Bie P (1980) Osmoreceptors, vasopressin and control of renal water excretion. Physiol Rev 60:4 961 – 1033
Biscoe JJ (1971) Carotid body: structure and function. Physiol Rev 51:437 – 491
Bloom SR, Edwards AV (1980) Effects of autonomic stimulation on the release of vasoactive intestinal peptide from the gastrointestinal tract in the calf. J Physiol (Lond) 299:437 – 452
Bonvallet M, Dell P, Hiebel G (1954) Tonus sympathique et activité électrique corticale. Electroenceph Clin Neurophysiol 6:119 – 144
Brobeck JR (1960) Regulation of feeding and drinking. Handbook of physiology. Neurophysiology, Sect 1, vol 2. American Physiological Society, Washington DC, pp 1197 – 1206
Brooks FP (1977) Nervous control of gastrointestinal function: neurophysiological considerations. Progr Gastroenterol 3:373 – 394
Brown AG, Iggo A (1967) A quantitative study of cutaneous receptors and afferent fibres in cat and rabbit. J Physiol (Lond) 193:707 – 733
Burgess PR, Perl ER (1973) Cutaneous mechanoreceptors and nociceptors. In: Iggo A (ed) Somatosensory system. Springer, Berlin, pp 29 – 78 (Handbook of sensory physiology, vol 2)
Burton H, Mc Farlane JJ (1973) The organization of the seventh lumbar spinal ganglion of the cat. J Comp Neurol 149:215 – 232
Cabanac M, Duclaux R (1973) Alliesthésie olfacto-gustative et prise alimentaire chez l'homme. J Physiol (Paris) 66:113 – 135
Cabanac M, Fantino M (1977) Origin of olfacto-gustatory alliesthesia: intestinal sensitivity to carbohydrate concentration. Physiol Behav 18:1039 – 1045
Cajal SR (1909) Histologie du système nerveux de l'homme et des vertébrés. Maloine, Paris, p 986
Calaresu FR, Kim P, Nakamura H, Sato A (1978) Electrophysiological characteristics of renorenal reflexes in the cat. J Physiol (Lond) 283:141 – 154
Clarke GD, Davison JS (1975) Tension receptors in the oesophagus and stomach of the rat. J Physiol (Lond) 244:41 – 42
Clarke GD, Davison JS (1978) Mucosal receptors in the gastric antrum and small intestine of the rat with afferent fibres in the cervical vagus. J Physiol (Lond) 284:55 – 67
Clerc N, Mei N (1981) Mise en évidence de mécanorécepteurs muqueux au niveau du sphincter inférieur de l'oesophage. Comparaison avec les mécanorécepteurs musculaires. CR Soc Biol (Paris) 175, 3:352 – 355
Clerc N, Mei N (1983) Vagal mechanoreceptors located in the lower oesophageal sphincter of the cat. J Physiol 336:487 – 498
Clifton GL, Coggeshall RE, Vance WH, Willis WD (1976) Receptive fields of unmyelinated ventral root afferent fibres in the cat. J Physiol (Lond) 256:573 – 600
Coggeshall RE, Coulter JD, Willis WD (1974) Unmyelinated axons in the ventral roots of the cat lumbosacral enlargement. J Com Neurol 153:39 – 58
Coleridge HM, Coleridge JCG, Banzett RB (1978) Effect of CO_2 on afferent vagal endings in the canine lung. Respir Physiol 34:135 – 151

Colindres RE, Spielman WS, Moss NG, Harrington WW, Gottschalk CW (1980) Functional evidence for renorenal reflexes in the rat. Am J Physiol 239:265 – 270
Crosby E, Humphrey T, Lauer E (1962) Correlative anatomy of the nervous system. Macmillan, New York, p 731
Crousillat J, Ranieri F (1980) Mécanorécepteurs splanchniques de la voie biliaire et de son péritoine. Exp Brain Res 40:146 – 153
Darcy B, Falempin M, Laplace JP, Rousseau JP (1979) Importance de la voie vagale sensitive: recherche d'une technique de déafférentation sélective chez le porc et le mouton. Ann Biol Anim Bioch Biophys 19:3B 881 – 888
Davison JS (1972) Response of single vagal afferent fibres to mechanical and chemical stimulation of the gastric and duodenal mucosa in cats. Q J Exp Physiol 57:405 – 416
Delpierre S, Grimaud C, Jammes Y, Mei N (1981) Changes in activity of vagal bronchopulmonary C-fibres by chemical and physical stimuli in cats. J Physiol (Lond) 316:61 – 74
Di Bella L, Tarozzi G, Rossi MT, Scalera G (1981) Effect of liver temperature increase on food intake. Physiol Behav 26:45 – 51
Droz B (1965) Rate of newly synthesized proteins in neurons. In: Leblond CP, Warrey KB (eds) The use of radioautography in investigating protein synthesis. Academic, New York, pp 159 – 175
Duclaux R, Mei N, Ranieri F (1976) Conduction velocity along the afferent vagal dendrites: a new type of fibre. J Physiol (Lond) 260:480 – 495
El Ouazzani T (1981) Contribution à l'étude de la sensibilité digestive. Approche histologique, histochimique et électrophysiologique chez le chat. Thèse Doctorat ès-Sciences Naturelles, p 269
El Ouazzani T, Mei N (1979) Mise en évidence électrophysiologique des thermorécepteurs vagaux dans la région gastrointestinale. Leur rôle dans la régulation de la motricité digestive. Exp Brain Res 34:419 – 434
El Ouazzani T, Mei N (1981) Acido- et glucorécepteurs vagaux de la région gastroduodénale. Exp Brain Res 42:442 – 452
El Quazzani T, Mei N (1982) Electrophysiologic properties and role of the vagal thermoreceptors of lower esophagus and stomach of cat. Gastroenterology 83:995 – 1002
Evans DHL, Murray JG (1954) Histological and functional studies on the fibre composition of the vagus nerve of the rabbit. J Anat 88:320 – 337
Falempin M (1981) Contribution à l'étude des afférences vagales digestives chez l'animal éveillé. Thèse Doctorat ès-Sciences Naturelles, p 195
Falempin M, Mei N, Rousseau JP (1978) Vagal mechanoreceptors of the inferior thoracic oesophagus, the lower oesophageal spincter and the stomach in the sheep. Pfluegers Arch 373:25 – 30
Fillenz M, Widdicombe JG (1972) Receptors of the lungs and airways. In: Neil E (ed) Enteroreceptors. Springer, Berlin, pp 81 – 112 (Handbook of sensory physiology, vol III/1)
Floyd K, Hick VE, Morrison JFB (1976) Mechanosensitive afferent units in the hypogastric nerve of the cat. J Physiol (Lond) 259:457 – 471
Floyd K, Hick VE, Koley J, Morrison JFB (1977) The effects of bradykinin on afferent units in intraabdominal sympathetic nerve trunks. Q J Exp Physiol 62:19 – 26
Fox KA, Kipp SC, Vanderweele DA (1976) Dietary self-selection following subdiaphragmatic vagotomy in the white rat. Am J Physiol 231:1790 – 1793
Freeman B, Rowe M (1981) The effect of sympathetic nerve stimulation of responses of cutaneous pacinian corpuscles in the cat. Neuroscience Lett 22:145 – 150
Gabella G (1976) Structure of the autonomic nervous system. Chapman and Hall, London, p 214
Gallego R, Eyzaguirre C (1978) Membrane and action potential characteristics of A and C nodose ganglion cells studied in whole ganglia and in tissue slices. J Neurophysiol 41:1217 – 1232

Gammon GD, Bronk DW (1935) The discharge of impulses from pacinian corpuscles in the mesentery and its relation to vascular changes. Am J Physiol 114:77–84
Garnier L (1968) Modifications EEG produites par la distension gastrique chez le chat. C R Soc Biol 162:2164–2168
Garnier L, Mei N (1982) Do true osmoreceptors exist at intestinal level? J Physiol (Lond) 327:97–98
Garnier L, Roman A, Pellegrino J (1965) Modifications de l'EEG du chat consécutives à la stimulation du nerf splanchnique. J Physiol (Paris) 57:245–246
Garcia M, Jordan D, Spyer KM (1979) Identification of aortic nerve cell bodies in the nodose ganglion of the rabbit and their central projection. J Physiol 290:23–24
Gasser HS (1955) Properties of dorsal root unmedullated fibers on the two sides of the ganglion. J Gen Physiol 38:704–728
Gaudin-Chazal G, Segu L, Seybritz M, Puizillout JJ (1981) Visualization of serotonin neurons in the nodose ganglia of the cat. An autoradiographic study. Neuroscience 6:1127–1137
Gonella J (1978) La motricité digestive et sa régulation nerveuse. J Physiol (Paris) 74:131–140
Gonella J, Gardette B (1974) Etude in vivo de la commande nerveuse extrinsèque parasympathique du côlon. J Physiol (Paris) 68:393–415
Gupta BN, Nier K, Hensel H (1979) Cold-sensitive afferents from the abdomen. Pfluegers Arch 380:203–204
Ha H (1970) Axonal bifurcation in the dorsal root ganglion of the cat: a light and electron microscopic study. J Comp Neurol 140:227–240
Haberich FJ (1968) Osmoreception in the portal circulation. Fed Proc 27:1137–1141
Hardcastle J, Hardcastle PT, Sanford PA (1978) Effect of actively transported hexoses on afferent nerve discharge from rat small intestine. J Physiol (Lond) 285:71–84
Harding R, Leek BF (1972a) Gastro-duodenal receptor responses to chemical and mechanical stimuli, investigated by a single fibre technique. J Physiol (Lond) 222:139–140P
Harding R, Leek BF (1972b) Rapidly adapting mechanoreceptors in the reticulorumen which respond to chemicals. J Physiol (Lond) 223:32–33
Hayashida Y, Koyano H, Eyzaguirre C (1980) An intracellular study of chemosensory fibers and endings. J Neurophysiol 44:1077–1088
Heinbecker P, O'Leary J (1933) The mammalian vagus nerve; a functional and histological study. Am J Physiol 106:623–646
Heyman SC, Neil E (1958) Reflexogenic areas of the cardiovascular system. Churchill, London, p 271
Howe A, Neil E (1972) Arterial chemoreceptors. In: Neil E (ed) Enteroreceptors. Springer, Berlin, pp 47–80 (Handbook of sensory physiology, vol III/1)
Hunt JN, Knox MT (1968) Regulation of gastric emptying. In: Handbook of physiology, sect 6, vol 4. American Physiological Society, Washington, DC, pp 1917–1935
Iggo A (1955) Tension receptors in the stomach and the urinary bladder. J Physiol (Lond) 128:593–607
Iggo A (1957a) Gastro-intestinal tension receptors with unmyelinated afferent fibres in the vagus. Q J Exp Physiol 42:130–141
Iggo A (1957b) Gastric mucosal chemoreceptors with vagal afferent fibres in the cat. Q J Exp Physiol 42:398–409
Ito M (1959) Analysis of potentials recorded intracellulary from the spinal ganglion. Cell Jpn J Physiol 9:33–41
Jaffe RA, Sampson SR (1976) Analysis of passive and active electrophysiologic properties of neurons in mammalian nodose ganglia maintained in vitro. J Neurophysiol 39:4 802–815
Jammes Y, Fornaris E, Mei N, Barrat E (1981) Sensory and motor composition of the bronchial vagal branches in cats. J Autonom Nerv Syst 5:165–176

Jeanningros R (1962) Vagal unitary responses to intestinal amino acid infusions in the anesthetized cat: a putative signal for protein-induced satiety. Physiol Behav 28:9 – 21
Jeanningros R, Mei N (1980) Données préliminaires sur la réponse des chémorécepteurs intestinaux aux acides aminés. Reprod Nutr Develop 20:1615 – 1619
Kametani H, Sato A, Sato Y, Simpson A (1979) Neural mechanisms of reflex facilitation and inhibition of gastric motility to stimulation of various skin areas in the rat. J Physiol (Lond) 294:407 – 418
Khabarova AY (1963) The afferent innervation of the heart. Pitman, London, p 175
Kiil J, Andersen D (1972) Evidence of a gastro-hepatic osmoregulation in humans and the influence of vagotomy on its activity. Scand J Gastroenterol 7:575 – 581
Kreulen DL, Szurszewski JH (1979) Nerve pathways in celiac plexus of the guinea pig. Am J Physiol 237:E90 – E97
La Barre J (1927) Sur l'augmentation de la teneur en insuline du sang veineux pancréatique après excitation du nerf vague. C R Soc Biol 96:193 – 196
Lasek RJ (1968) Axoplasmic transport in cat dorsal root ganglion cells: as studied with (^{3}H)-L-leucine. Brain Res 7:360 – 377
Leek BF (1969) Reticulo-ruminal mechanoreceptors in sheep. J Physiol (Lond) 202:585 – 609
Leek BF (1972) Abdominal visceral receptors. In: Neil E (ed) Enteroreceptors. Springer, Berlin, pp 113 – 160 (Handbook of sensory physiology, vol III/1)
Leek BF (1977) Abdominal and pelvic visceral receptors. Br Med Bull 33:163 – 168
Leitner JM, Perl ER (1964) Receptors supplied by spinal nerves which respond to cardiovascular changes and adrenaline. J Physiol (Lond) 175:254 – 274
Lieberman AR (1976) Sensory ganglia. In: Landon DN (ed) The peripheral nerve. Chapman and Hall, London, pp 188 – 278
Louis-Sylvestre J (1978) Relationship between two stages of prandial insulin release in rats. Am J Physiol 235:E103 – E111
Matsuda Y, Yoshida S, Yonezawa T (1978) Tetrodotoxin sensitivity and Ca component of action potentials or mouse dorsal root ganglion cells cultured in vitro. Brain Res 154:69 – 82
Mei N (1966) Existence d'une séparation anatomique des fibres vagales efférentes et afférentes au niveau du ganglion plexiforme du chat. J Physiol (Paris) 58:253 – 254
Mei N (1970a) Disposition anatomique et propriétés électrophysiologiques des neurones sensitifs vagaux chez le chat. Exp Brain Res 11:465 – 479
Mei N (1970b) Mécanorécepteurs vagaux digestifs chez le chat. Exp Brain Res 11: 502 – 514
Mei N (1976) Respiratory effects produced by the vagal afferents originating from the gastro-intestinal tract. In: Respiratory centres and afferent systems. INSERM 59:225 – 229
Mei N (1978) Vagal glucoreceptors in the small intestine of the cat. J Physiol (Lond) 282:485 – 506
Mei N (1981) La sensibilité viscérale. J Physiol (Paris) 77:597 – 612
Mei N, Roubien R (1970) Localisation des divers types de cellules dans le ganglion de Gasser du chat. Etude électrophysiologique par microélectrodes extracellulaires. J Physiol (Paris) 62:297 – 298
Mei N, Arlhac A, Boyer A (1981) Nervous regulation of insulin release by the intestinal vagal glucoreceptors. J Autonom Nerv Syst 41:351 – 363
Mei N, Condamin M, Boyer A (1980) The composition of the vagus nerve of the cat. Cell Tissue Res 209:423 – 431
Molhant M (1913) Le nerf vague. Etude anatomique et expérimentale. III. Les ganglions périphériques du vague. Névraxe 15:521 – 579
Morrison JBF (1973) Splanchnic slowly adapting mechanoreceptors with punctate receptive fields in the mesentery and gastrointestinal tract of the cat. J Physiol (Lond) 233:349 – 361

Neuhuber W, Nierderle B (1979) Spinal ganglion cells innervating the stomach of the rat as demonstrated by somatopetal transport of horseradish peroxidase (HRP). Anat Embryol 155:355 – 362

Newson B, Ahlman H, Dahlström A, Nyhus LM (1982) Ultrastructural observations in the rat ileal mucosa of possible epithelial "taste cells" and submucosal sensory neurons. Acta Physiol Scand 114:161 – 164

Nicolaidis S (1978) Rôle des réflexes anticipateurs oro-végétatifs dans la régulation hydrominérale et énergétique. J Physiol (Paris) 74:1 – 29

Niijima A (1969a) Afferent impulse discharges from glucoreceptors in the liver of the guinea pig. Ann NY Acad Sci 157:690 – 700

Niijima A (1969b) Afferent discharges from osmoreceptors in the liver of the guinea pig. Science NY 166:1519 – 1520

Niijima A (1971) Afferent discharges from arterial mechanoreceptors in the kidney of the rabbit. J Physiol (Lond) 219:477 – 485

Novin D (1976) Visceral mechanisms in the control of food intake. In: Nevin D, Wyrwicka W, Bray G (eds) Hunger: Basic mechanisms and clinical implications. Raven, New York, pp 357 – 367

Novin D, Sanderson J, Gonzales M (1979) Feeding nutrient infusions: effects of hypothalamic lesions and vagotomy. Physiol Behav 22:107 – 113

Paintal AS (1954) A study of gastric stretch receptors. Their role in the peripheral mechanism of satiation of hunger and thirst. J Physiol (Lond) 12:255 – 270

Paintal AS (1955) The study of ventricular pressure receptors and their role in the Bezold reflex. Q J Exp Physiol 40:348 – 363

Paintal AS (1957) Responses from mucosal mechanoreceptors in the small intestine of the cat. J Physiol (Lond) 139:353 – 368

Paintal AS (1963) Vagal afferent fibres. Erg Physiol 52:74 – 116

Paintal AS (1968) Some considerations relating to studies on chemoreceptor responses. In: Torrance RW (ed) Arterial chemoreceptors. Blackwell Scientific, Oxford, pp 253 – 262

Paintal AS (1972) Cardiovascular receptors. In: Neil E (ed) Enteroreceptors. Springer, Berlin, pp 1 – 45 (Handbook of sensory physiology, vol III/1)

Paintal AS (1973) Vagal sensory receptors and their reflex effects. Physiol Rev 53:159 – 226

Perrin J, Crousillat J, Mei N (1981) Assessment of true splanchnic glucoreceptors in the jejuno-ileum of the cat. Brain Res Bull 7:625 – 628

Portalier P, Vigier D (1979) Localization of aortic cells in the nodose ganglion by HRP retrograde transport in the cat. Neuroscience Lett 11:7 – 11

Powley TL, Weingarten HP, Opsahl CA (1976) The vagi and food intake. In: Brooks FP, Evers PN and Slack CB (eds) Nerves and the gut. Thorofare, New York, pp 459 – 479

Puizillout JJ, Foutz AS (1977) Characteristics of the experimental reflex sheep induced by vago-aortic nerve stimulation. Electroencephalogr Clin Neurophysiol 42:552 – 563

Ranieri F, Mei N, Crousillat J (1973) Les afférences splanchniques provenant des mécanorécepteurs gastrointestinaux et péritonéaux. Exp Brain Res 16:276 – 290

Ranson SW, Foley JO, Alpert CD (1933) Observations on the structure of the vagus nerve. Am J Anat 53:289 – 315

Rawson RO, Quick KP (1972) Localisation of intra-abdominal thermoreceptors in the ewe. J Physiol (Lond) 222:665 – 677

Rawson RO, Quick KP, Coughlin RF (1969) Thermo-regulatory responses to intra-abdominal heating of sheep. Sciences 165:919 – 920

Recordati GM, Moss NG, Waselkov L (1978) Renal chemoreceptors in the rat. Circ Res 43:534 – 543

Recordati GM, Moss NG, Genovesi S, Rogenes PT (1980) Renal receptors in the rat sensitive to chemical alterations of their environment. Circ Res 46:395 – 405

Reimann KA, Weaver LC, Meckler RLK, Oehl RS (1979) Organization of thoracic sympathetic afferent influences on renal nerve activity. Am J Physiol 237:H44 – H50

Reimann KA, Weaver LC (1980) Contrasting reflexes evoked by chemical activation of cardiac afferent nerves. Am J Physiol 239:H316–325

Riedel W (1976) Warmy receptors in the dorsal abdominal wall of the rabbit. Pfluegers Arch 361:205–206

Rignault D (1971) Séquelles générales et métaboliques de la chirurgie gastrique. Encyclopédie médico-chirurgicale 9029 D10:1–14

Rogers RC, Novin D (1980) The neurology of hepatic osmoregulation and its implications for studies of ingestive behavior. Brain Res Bull 5(4):189–194

Roman C (1966) Contrôle nerveux du péristaltisme oesophagien. J Physiol (Paris) 58:79–108

Roman C, Gonella J (1981) Extrinsic control of digestive tract motility. In: Johnson LR (ed) Physiology of the gastrointestinal tract, Raven, New York, pp 289–333

Rousseau JP (1970) Contribution à l'étude de la rumination et de l'éructation chez le mouton. Thèse Doctorat ès-Sciences Naturelles, Marseille, p 156

Roze C, Couturier D, Chariot J, Debray C (1977) Inhibition of gastric electrical and mechanical activity by intraduodenal agents in pigs and effects of vagotomy. Digestion 15:526–539

Ruckebusch Y (1961) Influx afferents d'origine péricardique dans les nerfs phréniques. C R Soc Biol 155:524–527

Sampson SR, Mills E (1970) Effects of sympathetic stimulation on discharges of aortic sinus baroreceptors. Am J Physiol 218:1650–1653

Sato A, Schmidt RF (1973) Somatosympathetic reflexes: afferent fibers, central pathways, discharge characteristics. Physiol Rev 53:916–947

Sato M, Austin G (1961) Intracellular potentials of mammalian dorsal root ganglion cells. J Neurophysiol 24:569–582

Sato Y, Terui N (1976) Changes in duodenal motility produced by nervous mechanical stimulation of the skin in rat. Neuroscience Lett 2:189–193

Sharma KN, Nasset ES (1962) Electrical activity in mesenteric nerves after perfusion of gut lumen. Am J Physiol 202:725–730

Singer VM, Solomon TE, Wood J, Grossman MI (1980) Latency of pancreatic enzyme response to intraduodenal stimulants. Am J Physiol 238:623–629

Stenquist B (1979) Studies on vagal activation of gastric acid and secretion in man. Acta Physiol Scand [Suppl] 465:1–31

Svaetichin G (1951) Analysis of action potentials from single spinal ganglion cells. Acta Physiol Scand 86:23–57

Svaetichin G (1957) Component analysis of action potentials from single neurons. Exp Cell Res [Suppl] 5:234–261

Szurszewski JH (1981) Physiology of mammalian prevertebral ganglia. Ann Rev Physiol 43:53–68

Temesreskasi D (1956) Vergleichend-histologische Untersuchungen an den Nervenelementen der Speiseröhre. Z Mikrosk Anat Forsch 62:661–672

Van Gehuchten A (1892a) Contribution à l'étude des ganglions cérébrospinaux. La cellule 8:209–231

Van Gehuchten A (1892b) Nouvelles recherches sur les ganglions cérébrospinaux. La Cellule 8:233–254

Widdicombe JG (1954a) Receptors in the trachea and bronchi of the cat. J Physiol (Lond) 123:55–70

Widdicombe JG (1954b) The site of pulmonary stretch receptors in the cat. J Physiol (Lond) 123:336–351

Woods SC, Porte D (1974) Neural control of endocrine pancreas. Physiol Rev 54:596–619

Yoshida S, Matsuda Y (1979) Studies on sensory neurons of the mouse with intracellular recording and horseradish peroxidase injection techniques. J Neurophysiol 42:1134–1145

Schematic Eye Models in Vertebrates

G. R. Martin

Department of Zoology and Comparative Physiology, University of Birmingham,
P.O. Box 363, Birmingham B15 2TT, England

1 Introduction and Brief Historical Perspective

The fascinating diversity of vertebrate ocular structure was first captured in a single work by Soemmering in 1818. Over a century and a half after their publication, Soemmering's engravings (Fig. 1) still raise a host of intriguing questions regarding how the vertebrate eye works as an optical system. Equally intriguing are the questions which are raised about the functions of the variety of eye shapes and sizes and the subsequent elucidation of general principles which may account for the diversity of eye structures.

The investigation of questions grouped around these two basic themes of mechanism and function are, in general, placed within quite different theoretical frame-

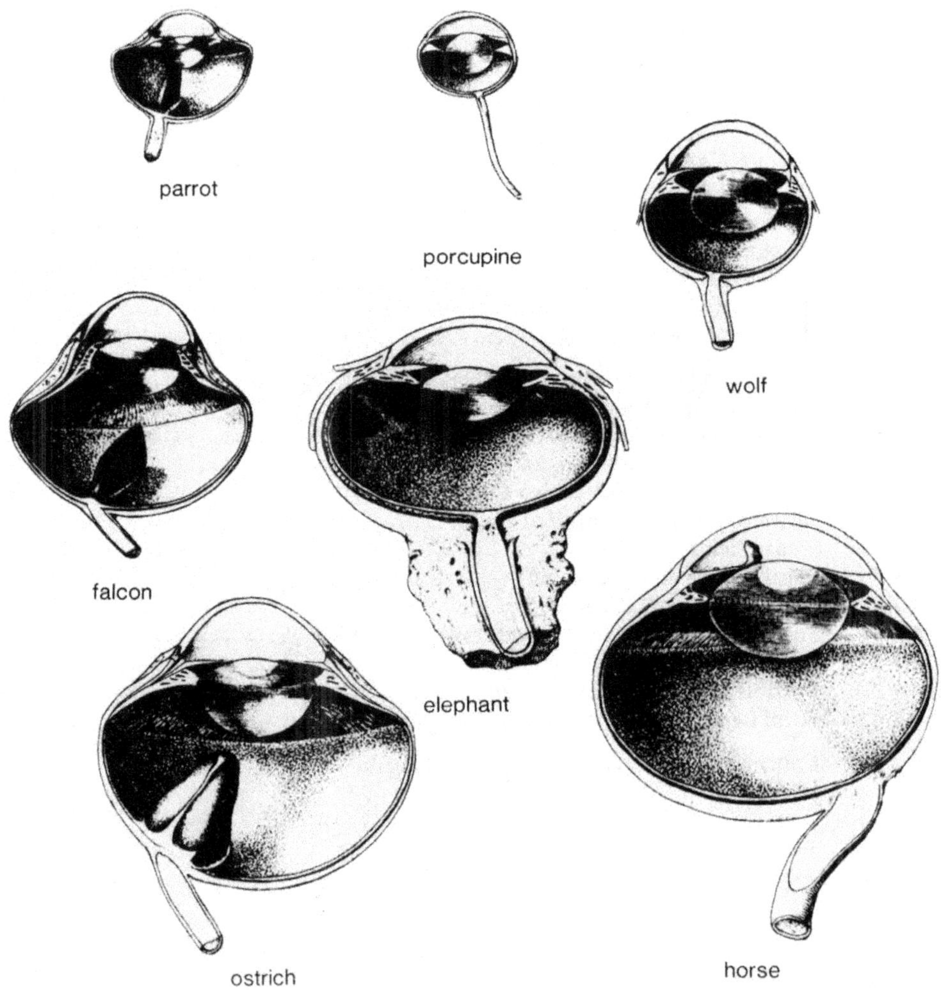

Fig. 1. A selection of Soemmering's (1818) engravings of avian and mammalian eyes showing the lower half of the hemisected left eye of each species. All drawings are to the same scale

works. However, in the case of the vertebrate eye, these two lines of questions come together in their common use of one particular tool – the schematic eye.

Put simply, the schematic eye is a brief mathematical model of the eye's optical system. It is usually regarded as representing an hypothetical, average eye for the species. The model is based upon the assumptions of geometrical optics and typically employs the paraxial approximations. It quantitatively describes certain key parameters of the optical system in question. Its value lies in the fact that these parameters can be put to a number of uses. They can be used for comparison between the actual performance of real eyes and that of an ideal system which the eye is thought to resemble. They can be used for intra- and interspecific comparisons of optical mechanisms, and in questions concerning the possible adaptive significance of different ocular structures, they can be used in comparisons between eye structure, visual performance and parameters of the natural visual environment.

In short, the value of the schematic eye is that it provides a quantitative basis for understanding the optics of the vertebrate eye, as opposed to explanations based upon anatomical descriptions and assumed optical properties.

Von Helmholtz (1909) considered Johannes Kepler the pioneer of physiological optics with his publication of *Dioptrice* in 1611, in which appeared the first description of how the eye formed an image on the retina. This description was a modification of an earlier one by Porta (1583) [quoted by Southall (1924)] in which the eye was likened to a camera obscura in which the image is formed in or on the lens rather than on the retina. Christian Huygens (1652) [quoted by Southall (1924)] was the first to describe and build a model of the eye, but it was not until 200 years later that the first schematic eye, that of man, was described (Listing 1845) [quoted in von Helmholtz (1909)]. This description followed the publication by Thomas Young in 1801 of his "On the mechanism of the eye" [Pumphrey (1961) considers the publication of this work to have been the true foundation of physiological optics] and the Gaussian simplifications for describing optical systems (Gauss 1841) upon which much of today's optical analysis is based.

Summarizing Listing's achievements in describing the optical system of the human eye in schematic terms, von Helmholtz (1909) concluded that, "There is no reason to doubt that Listing's model agrees about as well with the actual facts as can be expected with the wide range of variation that exists in individual eyes" (1924 translation, p. 95). In the context of interspecific comparisons and functional considerations, it can be argued that despite improvements in the precision and ease of optical measurements this conclusion is still as valid today as it was over a hundred years ago. However, workers subsequent to Listing, who have been concerned with providing increasingly comprehensive models of the optical mechanism of the eye, have proposed a number of alternative human schematic eye models. The one most commonly referred to is that of Gullstrand (1909). A number of so-called simplified schematic eyes (models based on three refracting surfaces; the cornea is reduced to a single surface) have also been proposed, e.g. von Helmholtz (1909), Tscherning (1898), Gullstrand (1909), as well as a number of 'reduced schematic eyes' (models based on only one refracting surface), e.g. Listing (1845) [quoted in von Helmholtz (1909)], Emsley (1948), Hughes (1979b). Although the reduced and simplified eyes have been proposed mainly for computational convenience, they owe their context to the attempts pioneered by Gullstrand (1909) to provide increasingly accurate models of the optics of the human eye. In recent years the attempt has been mainly to provide optical models of the lens and, in turn, the whole eye, which conform in part with the known anatomical and optical details of that structure, especially its continuous gradient of refractive index. Although this problem was first studied a hundred years ago by Matthiessen (1880, 1886) in the case of the radially symmetrical refractive index distribution of the spherical fish eye lens, it was not until 70 years later that the complete solution to even this somewhat restricted problem was first published (Fletcher et al. 1954). It seems that it has been necessary to await the advent of the modern computer before such a model could be applied to the more complex, gradient index biconvex lens of mammals (Nakao et al. 1968, 1970; Campbell and Hughes 1981). The theoretical importance of these gradient index lens models cannot be underestimated, since all vertebrate lenses studied to date, and even the lenses of the simple eyes of cephalopods (Land 1981) and a spider (Blest and Land 1977), are of the gradient index type.

It is surprising to find that a schematic eye for a mammalian species other than man was not published until as recently as 1958 when Herman described a schematic eye for the rat. Although Matthiessen (1886) and Hirschberg (1882) had described the dioptrics of eyes in fish and amphibia, the study of the cat (Vakkur and Bishop 1963) appears to be the first example where data derived from schematic eyes was used in an explicitly comparative manner to elucidate the possible adaptive significance of ocular structures. However, it should be noted that this was not the primary concern of the authors of that paper. Such a comparative exercise was building upon the traditions exemplified by the works of Rochon-Duvigneaud (1943), Walls (1942) and Duke-Elder (1958) in which the diversity of ocular structures is interpreted, within the context of the twin principles of phylogenetic relatedness and ecological convergence, as adaptation to different visual environments. The importance of the use of data from schematic eye models in such an exercise is that they add some quantitative rigour and hence admit the possibility of comparisons which are less speculative than those based solely upon anatomical data.

In addition to that of man, complete schematic eyes of varying degrees of sophistication have now been described for 12 vertebrate species (Table 1). These species represent but a very small sample of the wide diversity of vertebrate ocular structures which have been described anatomically. Also the list presents a far from comprehensive sample of ocular structures, even at the taxonomic level of the vertebrate class, and it is biased towards laboratory and domesticated species.

Although complete schematic eyes are available for only 13 vertebrate species, in recent years there have been a number of incomplete studies of the optics of certain vertebrate eyes. These have used techniques drawn from those employed in constructing schematic eyes. Typically such studies have been concerned with making restricted quantitative comparisons or elucidating the function of particular ocular structures, rather than with discerning general rules underlying the variety of vertebrate ocular structures. Studies of this kind include the following: the optics of the eyes of echolocating bats (Suthers and Wallis 1970); the so-called ramp retina in the eye of the horse (Sivak and Allen 1975) and the sting-ray (Sivak 1976b); the optical significance of the spectacle of the snake eye (Sivak

Table 1. Vertebrate species for which complete schematic eyes are available

Class	Species	
Osteichthyes	Goldfish, *Carassius auratus*	Charman and Tucker (1973)
Amphibia	Frog, *Rana esculenta*	Du Pont and De Groot (1976) Rivamonte (1977)
Reptilia	Turtle, *Pseudemys scripta elegans*	Northmore (1977)
	Gecko, *Gekko gekko*	Citron and Pinto (1973)
	Iguana, *Iguana iguana*	Citron and Pinto (1973)
Aves	Pigeon, *Columba livia*	Marshall et al. (1973)
	Owl, *Strix aluco*	Martin (1982)
Mammalia	Monkey, *Macaca mulatta*	Vakkur (1967) [quoted in Hughes (1977a)]
	Rabbit, *Oryctolagus cuniculus*	Hughes (1972)
	Rat, *Rattus norvegicus* (various laboratory breeds)	Hermann (1958) Block (1969) Massof and Chang (1972) Hughes (1979) Campbell and Hughes (1981)
	Cat, *Felis catus*	Vakkur and Bishop (1963)
	Mink, *Mustela vison*	Dunstone (personal communication)

1977), the 'flat' cornea of the penguin eye (Sivak 1976c; Sivak and Millodot 1977); the optics of the eye of the so-called four-eyed fish (*Anableps anableps*) (Sivak 1976a); and the refractive significance of the nictitating membrane in aquatic birds (Sivak et al. 1978).

2 The Description of Optical Systems

The purpose of most optical systems is to form images. In order to form a *perfect* image (by definition, one which is indistinguishable from the object except by reference to something other than the light used to form it) the optical system must fulfil two requirements: (1) it must bring all of the light collected from each independent object point together into an image point, and (2) it must also arrange these image points in an array which is geometrically similar to the object point array. Of course, images formed by actual optical systems are never quite perfect, and it is mainly with consideration of the imperfections which result from failure to fulfil the first requirement that optical science, including physiological optics, is most often concerned. However, the understanding of the important basic comparative features of all optical instruments, including eyes, is most readily achieved by temporarily neglecting these imperfections and describing the system in terms of just a few basic parameters.

Six parameters are typically used to describe optical systems. Depending upon the type of system, certain parameters are more salient when comparisons between individual systems are intended. The six parameters are: focal length, pupil aperture, *f*-number, magnification, field of view and depth of field. These are discussed in more detail in Sect. 4.1, with particular reference to what they reveal about the performance of, and design constraints on, vertebrate eyes. However, just two of these parameters, focal length and pupil aperture, are of primary importance, since they, in turn, define the limits of the other parameters.

In terms of function, optical systems are alloted to one of two main categories: *linear magnification systems,* whose function is to provide a real image of a selected object, and *angular magnification systems*, whose function is to provide angular magnification of a selected object. In the latter category are microscopes and telescopes, and they have but a single, fundamental parameter for their description and comparison: magnification. In the former group are such systems such as projectors, cameras and simple eyes. The convenient thing about these linear magnification systems is that in terms of the approximations of paraxial optics (see below), they can be considered as equivalent to a single, thin lens. Thus it is possible to describe the optical system of a vertebrate eye in the same shorthand terms as are applied to (say) a camera lens, i.e. the focal length of an equivalent thin lens, and its *f*-number [focal length (f)/entrance pupil diameter (A)]. In the case of the vertebrate eye, because the image is produced in a medium (vitreous humour or retinal tissue) of different refractive index from that of the object (air or water), the two focal lengths, anterior and posterior, are not equal. It is the anterior focal length (posterior nodal distance or PND) which is of prime importance, since it defines the size of the retinal image (see Fig. 2).

However, these shorthand encapsulations of optical properties provide only the basic starting point for describing optical performance. A complete description also requires details of image quality (see, for example, the reviews of Westheimer 1970, 1972) and optical performance outside of the paraxial region. Progress has begun in modelling the latter (e.g. Lotmar 1971; Pomerantzeff et al. 1971; Campbell and Hughes 1981), but there is an inherent problem in elaborating links between descriptions of an eye in terms of schematic optics and comprehensive descriptions of image quality. The former depends upon the assumptions of geometrical optics, while the latter requires analysis in terms of wave optics.

The schematic eye sets out to describe the positions of the six cardinal points of the eye's optical system – the *focal points, principal planes* and *nodal planes.* These points define completely the optical performance of the system, but only for the *paraxial case*, i.e. when light rays are close to the optic axis and make an angle of 10° or less with it. The position of the cardinal points can be determined by computation or graphical ray tracing, and once the cardinal points are known the basic lens equations can be applied. Image position and magnification for a given object can be determined, and for a given pupil size, image brightness and depth of focus can also be estimated. Application of these formulae can thus be used as the first test of the eye model proposed. Clearly, if even in the paraxial case the real eye can be shown not to behave in the way predicted by these formulae, then a revised model of the optical system is called for.

Outside of the paraxial region, the application of these simple formulae should be abandoned and graphical ray tracing adopted in order to make predictions of optical performance. However, for some purposes, the application of these shorthand lens equations can be used when rays outside of the paraxial definition are considered, but judgement of when these further approximations of actual performance can be tolerated must depend upon the intended use of the results. For example, such approximations may be acceptable where comparisons of light-gathering powers are being considered by comparison of *f*-numbers, but not where retinal magnification in the periphery of the visual field or the size of the visual field itself is under consideration.

When considering descriptions of the eye's optical system, it should be borne in mind that an important or perhaps the primary (Westheimer 1972, p 450) purpose of the vertebrate eye is to provide accurate localization of the sources of light reaching the animal. Thus it can be argued that the primary importance of the single optical system of the vertebrate eye (as opposed to the multiple optical systems of compound eyes) arises from its fundamental optical property of being able to reproduce accurately on a small, basically two-dimensional surface, the geometrical distribution, over a wide field, of objects at different distances from the animal. It is perhaps with this fundamental property that the description of the optical system of the vertebrate eye should really begin. It is a property which among the senses is unique to the vertebrate eye, but one which as the result of modern optical engineering is so commonplace today that its remarkable nature is readily overlooked.

3 Construction of the Schematic Eye

As a first approximation, the optical system of the vertebrate eye can be considered to consist of two thick lenses, the cornea and the crystalline lens, which are arranged symmetrically about a common axis. Constructing the schematic eye consists of determining the direction of the optic axis and computing the position of the cardinal points for each lens system in turn. The optical parameters of the two lens systems are then combined to determine a new set of cardinal points for the system as a whole. The construction of a schematic eye is portrayed diagramatically in Fig. 2.

The basic data required are the refractive indices of the different media involved and the positions and curvatures of the surfaces where the refractive index changes. There are a number of different techniques available for the collection of such basic data, and the formulae for converting these into descriptions of optical parameters are found in many sources (Emsley 1948; Davson 1962; Vakkur and Bishop 1963; Brown 1965; Westheimer 1972; Longhurst 1973; Fincham and Freeman 1980). Gullstrand (1909) and Hughes (1972) give computing schedules for constructing a complete schematic eye.

It is best if each separate item of data (curvature, position, refractive index) is collected independently, and this will always require access to freshly excised eyes. However, surface positions and curvatures can be determined in vivo using one measure to calculate the value of another. Typically measurements are taken from fresh material, but measurements from material recently fixed in glutaraldehyde (e.g. Marshall et al. 1973) or rapidly frozen (e.g. Citron and Pinto 1973; Northmore 1977) have been used in the construction of schematic eyes. Refrac-

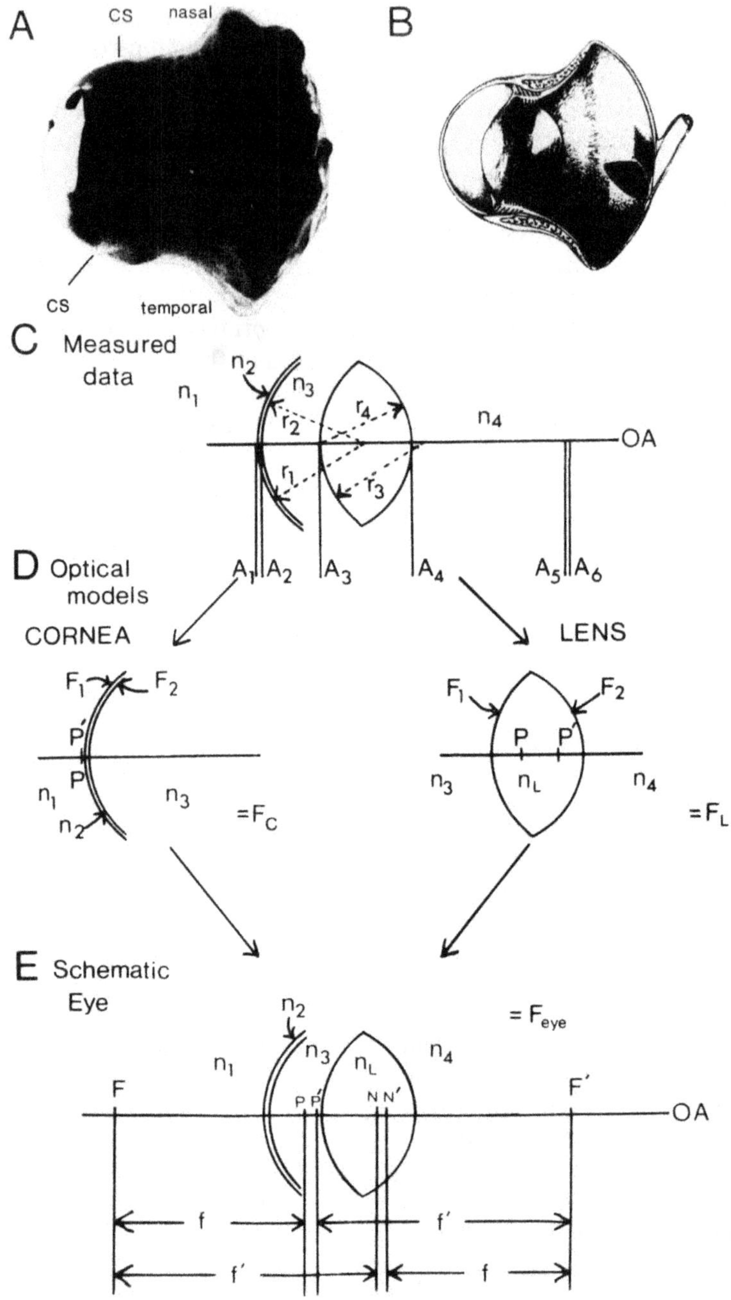

Fig. 2A – E. Diagrammatic construction of the schematic eye for an owl. **A** Photograph and **B** engraving of the eye under consideration; they both convey some information about the size, shape and general optical structure of the eye. It is of interest to compare them with the mathematical description of the same eye in **E. C** Measured data used to construct the schematic model. Positions of surfaces (A_1, A_2, A_3 *etc* ...) are measured along the optic axis (OA) from the corneal vertex A_1. Radii of curvature (r_1 and r_2, radii of corneal

tive index measures always require access to freshly dead material. Typically the refractive indices of the cornea, vitreous and aqueous are measured using an Abbe refractometer, though Marshall et al. (1973) used a Pulfrich refractometer. It is unnecessary to describe in detail the procedures for determing surface curvatures and positions. There are a number of procedures available, and most of these can be found described in the references to Table 1. Procedures which may be used with living eyes (their use being mainly restricted to man) are described in Sorsby et al. (1961), Davson (1962) and Westheimer (1972). In all cases, the most satisfactory techniques are those using photographs of fresh material, since they are rapid and reliable and allow leisurely analysis of results. A technique using X-rays to measure the axial length of the eye in vivo was described by Rushton (1938), but this is really applicable only to human subjects, since it requires the subject to report on a visual sensation. Jansson (1963) has reported an in vivo technique for measuring the positions of the refracting surfaces and axial length using an ultrasound echo-impulse method. The technique's main application would seem to be in clinical procedures, though it could be used with non-human subjects.

Fig. 2A – E. (*continued*) anterior and posterior surfaces, r_3 and r_4, radii of lens anterior and posterior surfaces) are measured from points on the optic axis, i.e. symmetry about the optic axis is assumed. The refractive indices n_1 (air), n_2 (cornea), n_3 (aqueous humour), n_4 (vitreous humour) are measured with a refractometer. Pupil apertures are usually measured by photographing the pupil along the direction of the optic axis in air (entrance pupil) and with the eye immersed in saline (real pupil) to effectively abolish the refractive power of the cornea. Other data which may be taken are the diameter of the corneo-scleral junction (*CS* in **A**), the maximum chord of curvature of the anterior and posterior lens surfaces, the thickness of the eye wall (A_5A_6) and the approximate radius and diameter of retinal surface. **D** Optical models of the cornea and lens considered separately. The surface powers F_1, F_2 are calculated from known refractive indices and their separations, A_1A_2 and A_3A_4. These are used to determine the total power of the cornea and the positions of the two principal points, PP'. In the case of the cornea, P and P' are practically coincident and are almost at A_1. The same procedure is used for the lens, but the lens assumed homogeneous refractive index (n_L) has to be determined by successive approximations [see, for example, Vakkur and Bishop (1963) for a description of one such technique]. **E** The final schematic eye model. The total power (F_{eye}) and the positions of the six cardinal points of the system are calculated from the optical models of the cornea and lens and their known separations. F and F' are the first and second focal points of the eye; F is the object point which would produce an infinitely distant image, F' the image point for an infinitely distant object (if F' coincides with the retina, then the eye model is emmetropic; if F' is placed anterior or posterior to the retina, the schematic eye is myopic or hyperopic, respectively). The anterior and posterior focal lengths and their relation to each other are indicated in the diagram. P and P' are the principal points which define the positions along the optic axis of the principal planes (planes of unit transverse magnification). N and N' are the nodal points, and these define the positions along the optic axis of the nodal planes (planes of unit angular magnification). The full description of the schematic eye includes (a) the positions of each surface and its curvature, (b) the positions of the six cardinal points, (c) the refractive indices of the various structures, and (d) the surface and total powers of the lens and cornea and the total power of the whole eye. Focal lengths are derived from the separation of the cardinal points, as in the diagram. A full description of the eye will also include data on the maximum aperture of the real, entrance, and exit pupils. f [anterior (first) focal length = posterior nodal distance (*PND*)] = n_1/F_{eye}; f' [posterior (second) focal length] = n_4/F_{eye}

3.1 Assumptions and Approximations

In addition to the assumptions of geometrical optics, the construction of the schematic eye always involves the use of a number of additional assumptions and approximations. These vary in their importance, and the ones which are common to most eye models are discussed below. A particular model may be based on only some of these assumptions, or it may employ assumptions in addition to those listed here.

3.1.1 The Optic Axis

That the optical elements (lens and cornea) are placed symmetrically about a common axis, and hence a single optic axis actually exists, is an erroneous assumption and was recognized as such by von Helmholtz (1909). However, an approximate optic axis can be easily discerned in all eyes by the simple technique of determining the locus at which the Purkinje images of a single light source are superimposed, or the locus at which the Purkinje images of two light sources are aligned (Westheimer 1972).

3.1.2 The Lens

The principal stumbling block in constructing a schematic eye is the description of the lens. With few exceptions (Pomerantzeff et al. 1971; Nakao et al. 1968, 1970; Campbell and Hughes 1981), all schematic eye models have been founded upon one of two explicit, but erroneous, assumptions regarding the lens. These are either (1) that the lens is of homogeneous refractive index, or (2) that it consists of just two refractive indices, a spherical core of one index and an outer layer of a lower index. As stated above, all vertebrate lenses studied to date have

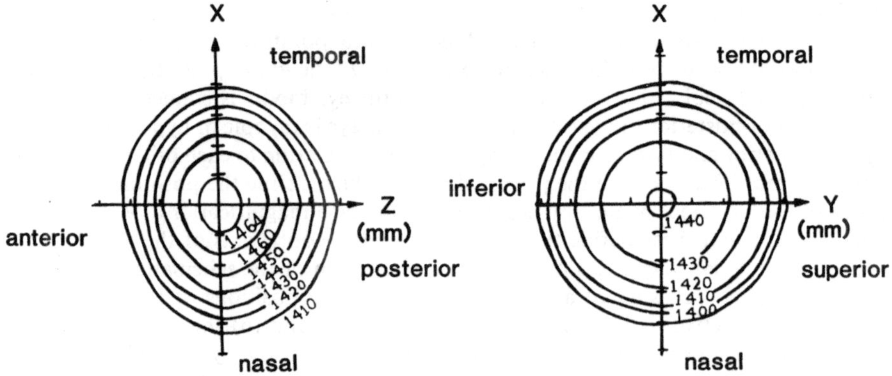

Fig. 3. Isoindicial contours of the rabbit lens in two perpendicular planes. The z axis is equivalent to the direction of the optic axis (Nakao et al. 1968)

been found to possess a continuous gradient of refractive index which decreases from the lens core to the periphery.
The function of a refractive index profile within the lens as a means of increasing lens power and reducing spherical aberration is theoretically well understood (Fletcher et al. 1954). However, the exact profile of the refractive index gradient is a complex three-dimensional function, and data for man, cat, monkey and rabbit (Nakao et al. 1968, 1969; Philipson 1969) show both intra- and interspecific variations in refractive index gradients (Fig. 3). Matthiessen (1866), Nakao et al. (1968), Pomerantzeff et al. (1971) and Campbell and Hughes (1981) have all studied the optical performance of such refractive index gradient lenses. Unfortunately, because of the complexity of the measures and computational techniques used, and because of intra- and interspecific variations of gradients, the possibility of general solutions, or of solutions which permit rapid interspecific comparisons of optical mechanisms, seems remote at this time. However, what all of these models do show most clearly is the exquisite subtlety of the optical design of the vertebrate lens, an optical design which is far superior to that presently seen in any man-made system.

3.1.3 Surface Curvature

The simplest procedure in constructing the schematic eye models is to assume that the refracting surfaces are both spherical and symmetrical about the optic axis. This leads to ease of measurement and computational simplicity. It is however worth noting the deviation of actual surface shapes from this ideal. While assuming sphericity does not detract seriously from the utility of schematic eyes for modelling performance within the paraxial region, consideration of the deviations of the actual surfaces from this ideal does reinforce the need for models which can deal with optics in the visual periphery and, hence, adequately model the known chromatic and monochromatic aberrations of the eye's peripheral optics. However, such an exercise should perhaps be embarked upon with caution. Westheimer (1972) has remarked that, "Detailed theoretical analysis of the monochromatic aberrations of eyes is not a rewarding study because individual variations are large enough to render any general conclusions invalid" (p 469).

3.1.3.1 Cornea. Bonnet (1964) proposed that the corneal curvature of the human eye can be described by the expression, $\log_{10}\beta = 3(\alpha - 1)$, where α and β are as defined in Fig. 4A. The actual corneal profile derived from this expression was used in a schematic eye model by Lotmar (1971) and Lotmar and Lotmar (1974), and is shown in Fig. 4B, where it is compared with the profiles of a circle, ellipse and parabola. The x axis can be taken as the optic axis, and it is clear that for rays close to that axis the optical properties of corneae with curvatures the same as these different surfaces will be virtually identical, and certainly well within the range of individual variation. The assumption that the corneal profile is radially symmetrical about the optic axis is also in error (Lopping and Weale 1965), though the optical significance of this asymmetry is unassessed.

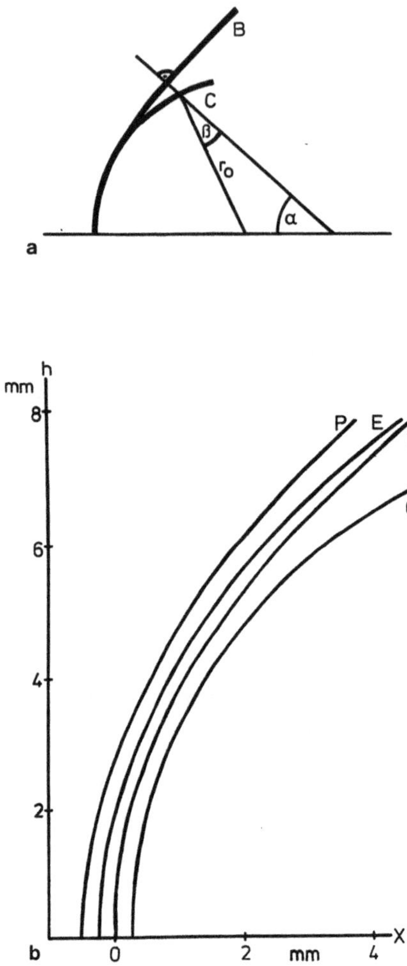

Fig. 4. a Definition of the angles used in Bonnet's (1964) formula, $\log_{10}\beta = k\alpha + b$ which describes the curvature of the cornea. *B* is the corneal profile, *C* an osculating circle. **b** Comparison between Bonnet's corneal profile curve (*B*) and three other curves: an ellipse (*E*); a parabola (*P*) and an osculating circle (*C*) (Lotmar 1971)

3.1.3.2 Lens. In man, neither the anterior nor the posterior surface of the lens is spherical, and this conclusion can probably be extended to all mammals and birds and also to certain fish (e.g. Sivak 1976a). A study of 60 pairs of human lenses led Howcroft and Parker (1977) to conclude that the anterior surface is typically hyperbolic, while the posterior surface is parabolic. To date, a model of the eye has not been proposed which takes simultaneous account of both the aspheric surfaces and the refractive index gradient of the lens. However, it should be noted that all of the assumptions described above progressively decrease the validity of the eye model as one considers optical performance further from the paraxial region. Within the assumptions of paraxial optics, even homogeneous lens models employing spherical surfaces would seem to provide a good approximation of actual average optical performance, given the individual variability of eye structures.

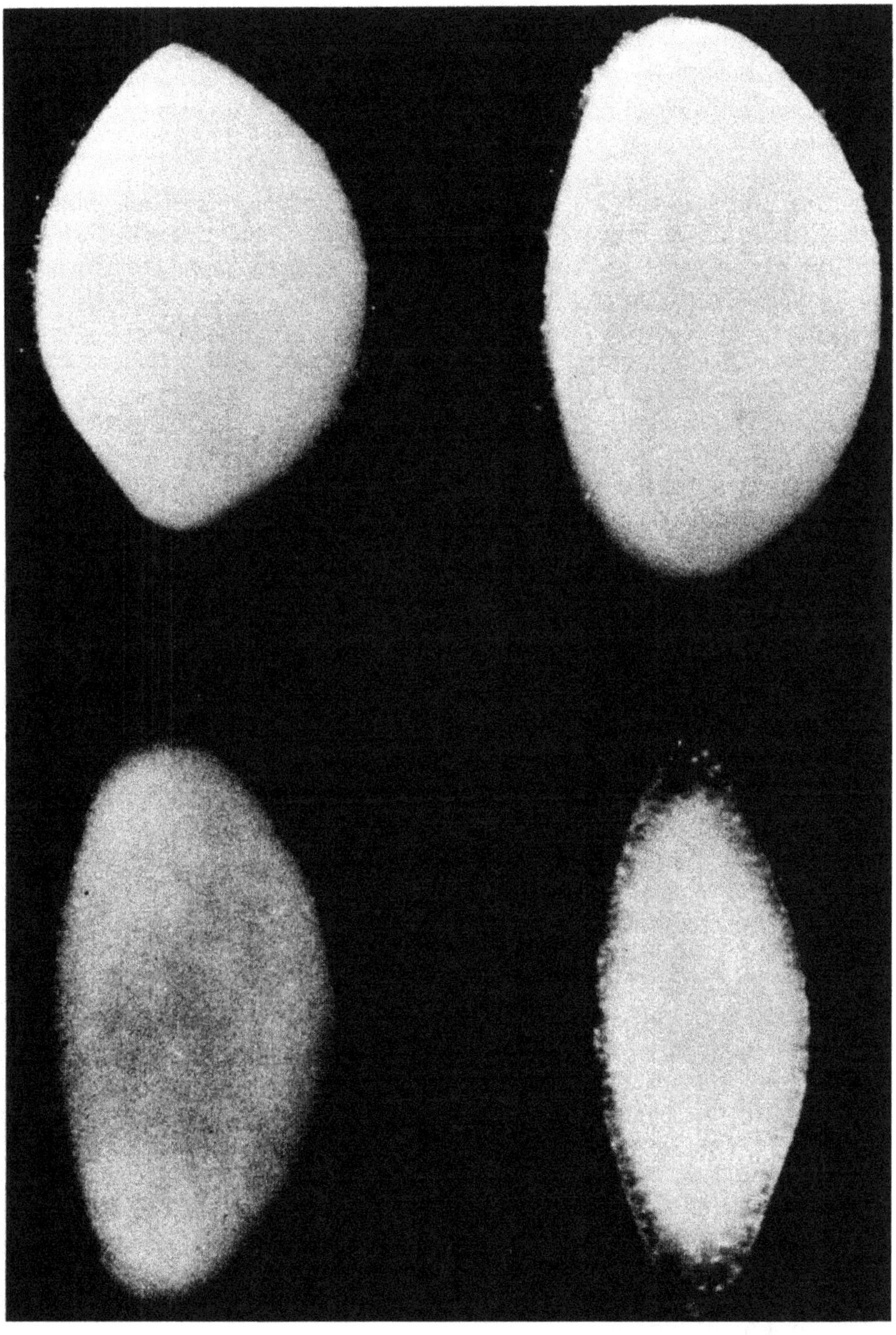

Fig. 5. Four profiles (frozen sections) of the human lens which illustrate the variation in shape recorded in a sample of 60 pairs of eyes by Howcroft and Parker (1977)

3.1.4 Individual Variation in Optical Parameters

Howcroft and Parker (1979) recorded a large variation in surface curvature and lens thickness in their sample of human lenses (Fig. 5). They also demonstrated a significant difference in these parameters with age, a feature also noted by Weeker et al. (1975).

Two extensive studies have investigated the variation found in a number of optical parameters in man (Sorsby et al. 1957, 1961). From a study of 341 eyes it was found that amongst those judged to be emmetropic (i.e. the image of a distant object is brought to a focus at the retina), axial eye length varied between 21 – 26 mm, corneal power varied between 38 – 48 dioptres (D) and lens power between 17 – 26 D. The range of total eye power amongst emmetropes was from 54.6 D to 66.3 D, equivalent to a difference in anterior focal length of from 18.3 mm to 15.1 mm. (Gullstrand's schematic values of total eye power = 58.64 D; anterior focal length = 17.055 mm.)

In the emmetropic eye there is a high correlation of axial length with both the corneal and the lens powers. While the various components of the eye's optical system are normally distributed in the population, Sorsby et al. (1957) conclude that, "The emmetropic eye is not the end result of a haphazard combination of variable components, but a co-ordinated organ. The individual emmetropic eye ... is not accidental" (p 39). Rather, it is the result of a highly co-ordinated development which would seem to depend upon normal visual experience. Certainly, in the case of the developing chick eye, only minor alteration of visual experience produces marked anomalous development in individual components (Wallman et al. 1978). In the human eye, refractive errors are the result of anomalous components (cornea and lens powers, depth of anterior chamber, axial length of eye) whose values fall outside the range of normal variation (Sorsby et al. 1957).

These measures of individual variation highlight the problem of defining the 'average' eye and its components, and add justification to the conclusion of von Helmholtz (1909) concerning the representativeness of Listing's schematic eye (Sect. 1). The main feature which average human eyes have in common is their emmetropia; eyes of this refractive state show a large variation in the optical properties of individual components. This variation immediately puts a limitation on the subtlety of interspecific differences in optical properties which can be sensibly discussed, as well as a limit on the degree of correlation which may be expected between parameters of the schematic eye and average visual performance (Sect. 4.1). Certainly, these results do suggest that schematic eyes based upon average data from a large sample of eyes are no more useful for answering the general questions posed in the first paragraph of this chapter than are schematic eyes based upon only a small sample or even upon individual cases.

3.1.5 Position of the Image

The exact origin of the retinoscopic reflex has been widely discussed (Glickstein and Millodot 1970; Millodot 1972; Charman and Jennings 1976; Millodot and Sivak 1978; Nuboer and Van Gendersen-Takken 1978; Bobier and Sivak 1978; Hughes 1977b, 1979c), but a consensus does not yet appear to have been reached, though Hughes (1977b) argues strongly in support of Glickstein and Millodot's (1970) hypothesis that the reflex originates at the retinal-vitreal boundary. The problem is important for the construction of schematic eyes of high accuracy which seek to model the exact position of the retina with regard to the refractive state of the eye (e.g. Hughes 1979a). The dimensions involved are small and their measurement becomes increasingly critical as eye size decreases. For example, in the human eye (axial length = 24.0 mm) a change of 1 D in the refractive state of the eye would be equivalent to an axial shift in image position of about 380 μm, while in the smaller rat eye (axial length = 6.3 mm), such a change in refractive state shifts the image position by only 15 μm. In view of these problems, coupled with the inherent variability of both ocular structures and measures of refractive state (Sorsby et al. 1957), it would seem satisfactory for many uses of the schematic eye simply to assume that the eye is emmetropic and position the retina accordingly (e.g. Martin 1982) or at least position the retina assuming that an emmetropic image falls at the vitreal surface of the retina (e.g. Citron and Pinto 1973).

4 Analysis and Interpretation of Schematic Eye Models

Figure 6 shows the essential features of the schematic eye models of six species. Each eye represents a unique set of properties which can be analysed in detail to answer specific questions about that particular species' optical system, and how results obtained in a specific experimental situation might be influenced. This approach has been most successfully pursued in the case of the rat (Hughes 1977b, 1979a) and cat (Vakkur and Bishop 1963) eyes where schematic eye data have been used in the design and interpretation of neurophysiological experiments.

Another value of the eye models emerges when they are viewed in a comparative context. Because of the many assumptions and approximations underlying schematic eyes (Sect. 3), such comparative analysis is most sensibly restricted to the principal descriptive parameters of linear magnification systems, focal length and *f*-number and to discussion of the principal factors which may account for the diversity of ocular design.

Such comparative analyses have their principal value in the task of interpreting the diversity of ocular structures as adaptations to different visual environments. However, in this latter task, in order to produce hypotheses which are of both wide generality and sufficiently well formulated to be readily testable, a broad comparative context is required. Unfortunately, a sufficiently wide diversity of eyes have not been fully described schematically to satisfactorily fulfil this requirement, and presently described hypotheses must be regarded as tentative at best.

4.1 Schematic Optics and Visual Performance

The two fundamental parameters of the schematic eye models, anterior focal length (or PND) and minimum *f*-number, are of importance in discussions under the present heading. Data on maximum pupil diameter is not available for all the

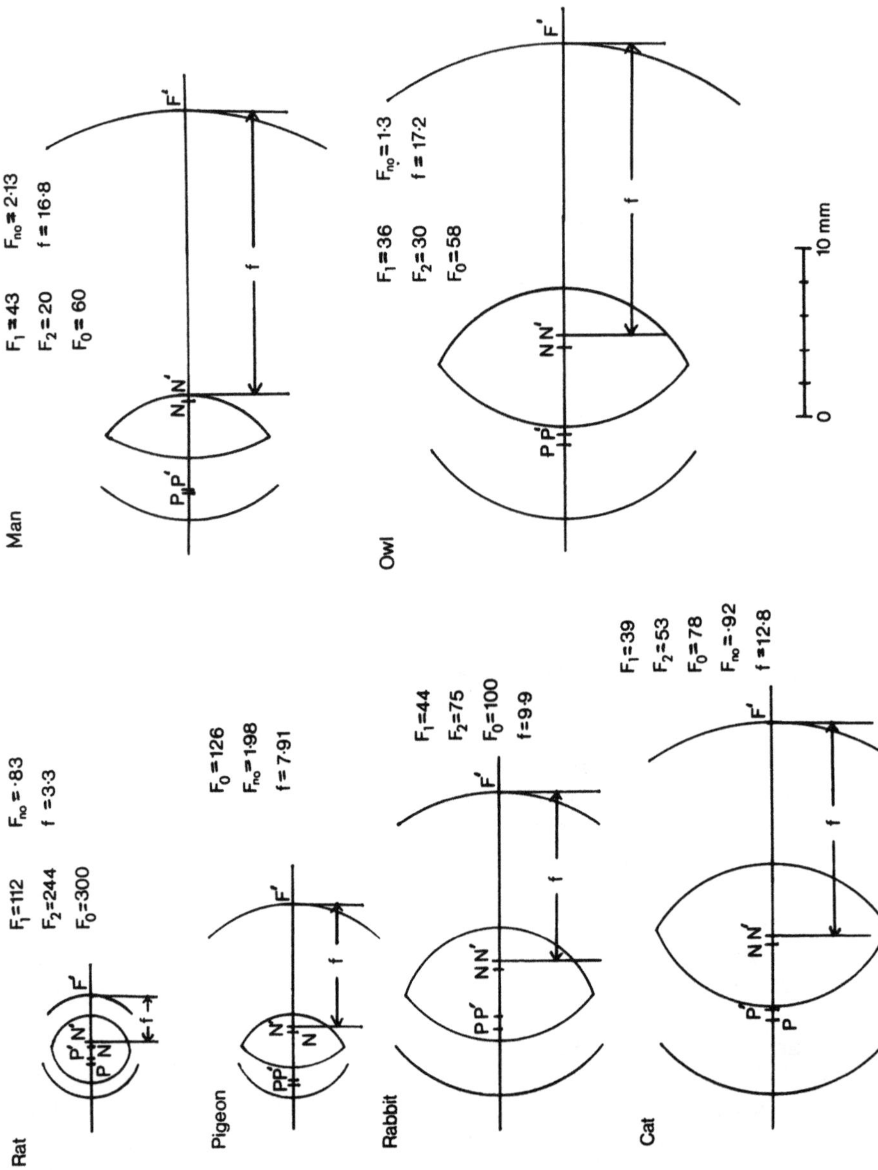

Fig. 6. Scale drawings of the schematic eyes of six species. F_1, F_2, F_0 are the refractive powers (in dioptres) of cornea, lens and whole eye, respectively. F_{no} is the minimum *f*-number (maximum pupil diameter) for the eye (F_{no} data from Table 2), *f* is the anterior focal length (PND) in mm. The principal points (P, P'), nodal points (N, N') and the second focal point (F') are shown. The retina is drawn passing through the second focal point. Rat (Hughes 1979a), pigeon (Marshall et al. 1973), rabbit (Hughes 1972), cat (Vakkur and Bishop 1963), man (Gullstrand 1909), owl (*Strix aluco*) (Martin 1982)

Table 2. Certain eye dimensions and relative image illumination in the eyes of man, cat, owl, pigeon and rat (Martin 1982)

Species	Maximum entrance pupil diameter (A) mm	Pupil area (mm^2)	Anterior focal length (f) mm	f-Number (f/A)	(f-Number)^2	Retinal image illumination relative to man			
						Point source		Extended source	
						×	$\log_{10}$	×	$\log_{10}$
Man	8.0[a]	50.27	17.06[b]	2.13	4.54	–	–	–	–
Cat	14.0[a]	153.9	12.82[a]	0.92	0.85	3.1	+0.49	5.4	+0.73
Owl	13.3	138.9	17.24	1.30	1.69	2.8	+0.44	2.7	+0.43
Pigeon	4.0[c]	12.57	7.91[d]	1.98	3.92	0.25	–0.60	1.2	+0.06
Rat	4.0[e]	12.57	3.33[e]	0.83	0.69	0.25	–0.60	6.6	+0.82

[a] Vakkur and Bishop (1963)
[b] Hughes (1977, Table 1)
[c] Personal observation
[d] Marshall et al. (1973)
[e] Hughes (1979)

species whose eyes have been described schematically. The relevent data for the present discussion are shown in Table 2.

Since the real pupil in the vertebrate eye is placed between the two optical elements (cornea and lens), it is necessary to distinguish between the *entrance pupil,* which is the magnified image of the real pupil formed in object space by the cornea (this is the pupil which we see when looking into the eye), and the *exit pupil,* which is the image of the real pupil formed in image space by the lens. Because the iris is in contact with the anterior surface of the lens, the real pupil not only changes in diameter but also changes in position. Hence, there are a set of complex relationships between entrance, exit and real pupil apertures and positions for a given eye. These have been analysed in detail by Vakkur and Bishop (1963) and Hughes (1979a) for the cat and rat eyes, respectively, using schematic eye data. Fortunately for most comparative purposes, it is only the maximum entrance and exit pupils which are important, and the difference in their positions may be ignored for all but the most comprehensive of analyses. Since most attention is devoted to minimum f-number, data from calibrated high-speed flash photographs of fully dilated pupils in the intact eye are sufficient for most analyses.

4.1.1 Light Gathering Power and Absolute Sensitivity

When an eye views an extended source of light, the brightness of the retinal image is proportional to the f-number of the eye, but when it views a point source of light, image brightness is simply proportional to pupil area (Kirschfeld 1974; Land 1981). These simple relationships can be used in the comparative analysis of schematic eye data, and they become particularly useful if psychophysical data on the visual performance, especially absolute visual threshold, of the same species are also available (Martin 1982).

'Nocturnal-diurnal' is a classificatory distinction which is used in the discussion of many aspects of anatomy, physiology and behaviour, and especially in discussion of eye structures and visual performance. Discussions of the characteristics thought typical of diurnal and nocturnal eyes are found in many places [e.g. Walls (1942), Duke-Elder (1958), Hughes (1977a)], with the human eye generally allocated to the diurnal category. However, it would seem that these analyses have been conducted in the absence of a rigorous sensory-ecological analysis of the differences between the nocturnal and diurnal environments, as well as a generally loose understanding of the nocturnal-diurnal distinction from the behavioural standpoint.

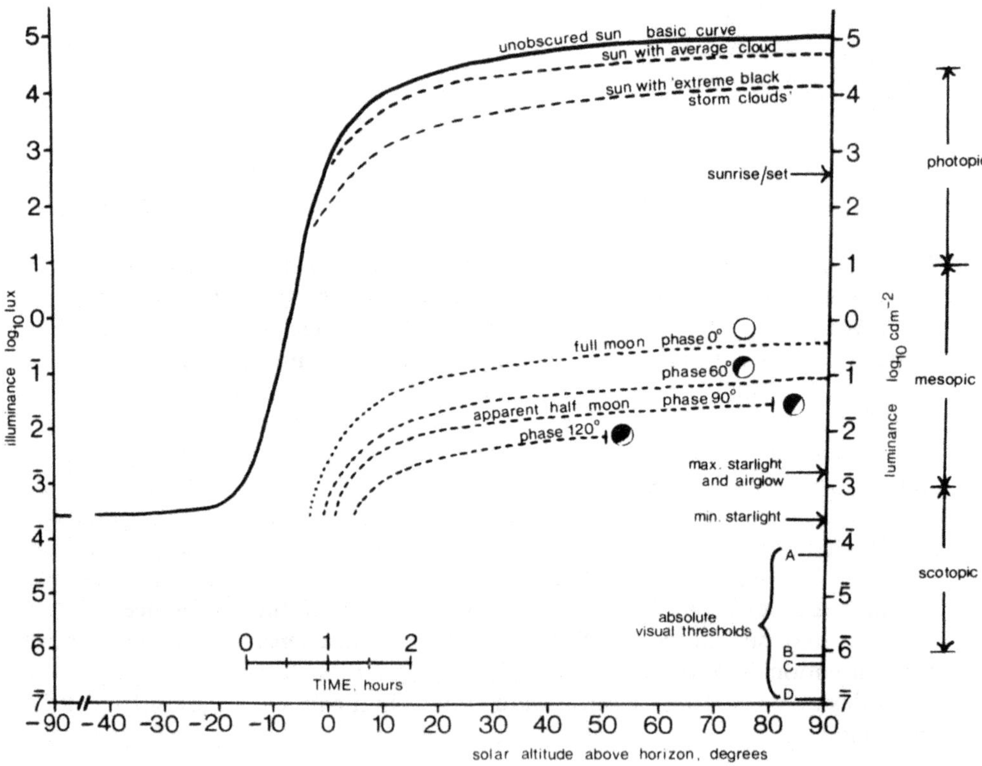

Fig. 7. Naturally occurring illumination levels and the luminance level which such illumination would produce on a highly reflecting and diffusing surface, such as white paper. Illumination levels due to the sun as a function of altitude, the moon as a function of altitude and phase, and starlight are indicated. The amount of attenuation produced by cloud cover is also shown. The basic sun curve is for latitude 50° at the time of the summer solstice. The absolute visual thresholds are for *A* (pigeon), *B* (man), *C* (owl), *D* (cat) (see Table 3). ***Photopic, mesopic*** and ***scotopic*** luminance ranges are from Pirenne 1962, p 67) [Based upon the Natural Illumination Charts of the US Navy (1952)]

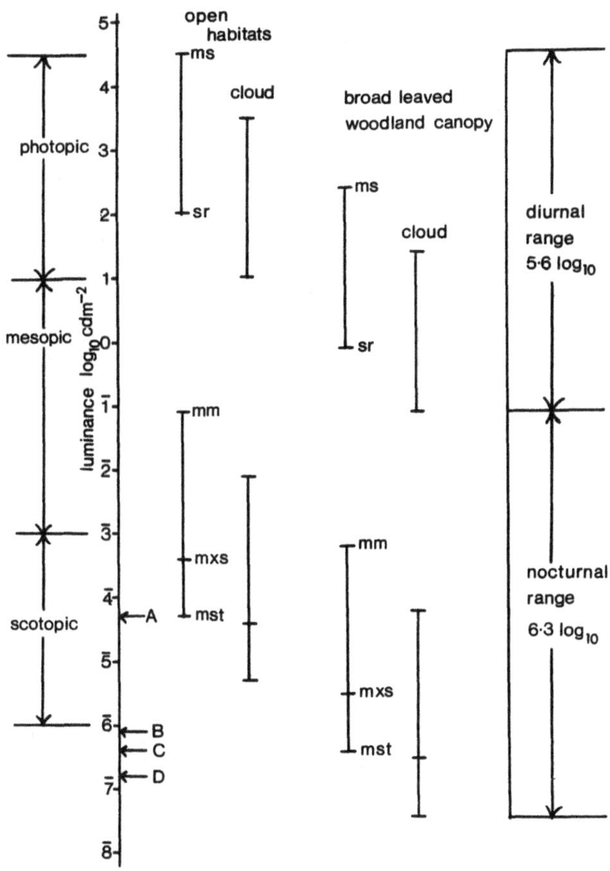

Fig. 8. Naturally occurring luminance levels of a grass or leaf litter substrate during day-time and night-time. Each *vertical bar* indicates the maximum likely range of luminance which can be experienced in the habitat indicated. Absolute luminance ranges are from the Natural Illumination Charts of the US Navy (1952) for latitude 50° at the time of the summer solstice (Fig. 7). Daytime levels show the luminance range from sunrise (*sr*) to maximum midday sun (*ms*). Night-time (sun altitude >8° below the horizon) luminance levels show the range from maximum moonlight (*mm*) to minimum starlight (*mst*). Maximum starlight is also indicated (*mxs*). Luminance levels experienced under maximum cloud cover [attenuation 1.0 $\log_{10}$ units, Taylor and Kerr (1941)] are also shown. The reflectance of grass or leaf litter is taken to equal 0.6 $\log_{10}$ (personal observation using a U.D.T. 80× optometer showed fresh grass and leaf litter to be variable in reflectance but with a mean of this magnitude). The mean attenuation produced by a broad-leaved tree canopy in full leaf is taken to be 2.1 $\log_{10}$ (Federer and Tanner 1966). The total diurnal luminance range of 5.6 $\log_{10}$ units ($4.5-\bar{2}.9$ log cdm^{-2}) is the range which may be experienced between maximum sunlight without cloud cover in an open habitat and at sunrise with cloud cover beneath the woodland canopy. The total nocturnal range of 6.3 $\log_{10}$ units ($\bar{2}.9-\bar{8}.6$ $\log_{10}$ cdm^{-2}) is the range from maximum moonlight without cloud in an open habitat to minimum starlight with cloud beneath the woodland canopy. Definition of the *photopic, mesopic* and *scotopic* ranges are from Pirenne (1962, p. 67) and points *A, B, C* and *D* indicate the mean behaviourally determined absolute visual threshold of pigeon, man, owl and cat respectively (see Table 3). From Martin (1982)

One context in which to view this dichotomy, from the point of view of visual science, is the total range of naturally occurring luminance levels and their degree of variability experienced within the terrestrial environment. The basic curve of Fig. 7 shows the naturally occurring illumination and the resulting luminance level of a highly reflecting surface (such as white paper) placed in the open, which occur during a 24-h period in an equatorial or temperate latitude during the summer months. The speed of change from night-time to daytime luminance levels varies with latitude and time of year and is most rapid at the equator; it decreases with increasing latitude until at the poles there is no discernable 24-h cycle, natural luminance simply increasing or decreasing from day to day, depending upon the season. Another change which occurs with increased latitude is a variation in the asymptotic daytime and night-time luminance values, such that at higher latitude, days are considerably less bright and nights less dark than they are at the equator. Superimposed on these regular and predictable, seasonal and latitudinal differences in luminance level throughout the 24 h are factors such as cloud cover and, during night-time, the presence of stars and the moon at different altitudes and phases. This data can be combined with known differences in the reflectance of different natural substrates and data on the light attenuation of natural tree canopies to give predictions of substrate luminance likely to be experienced in natural habitats. Figure 8 indicates that within temperate latitudes the maximum luminance range of a particular substrate or object which is likely to be experienced is 5.6 $\log_{10}$ units ($\times$ 400 000) between the hours of sunrise and sunset, and 6.3 $\log_{10}$ units ($\times$ 2 000 000) during the night, but with practically no overlap in the absolute extent of these ranges. Figure 8 also indicates that under variable cloud conditions an animal may experience a maximum range of about 3.1 log units ($\times$ 1200) in the luminance of a given object or substrate when moving between the open and a closed canopy, broad leaved woodland in full leaf, whether by day or night.

It is in the context of the kinds of luminance ranges presented in Fig. 8 that the light-gathering power and absolute sensitivity of vertebrate eyes is perhaps best assessed.

Table 3. Retinal image illumination for an extended light source and mean absolute visual threshold in cat, owl and pigeon, expressed relative to those of man[a]

Species	(f-Number)^2	Image illumination relative to man ($\log_{10}$)	Mean behaviourally determined absolute visual threshold relative to man ($\log_{10}$)
Man	4.54	–	–
Cat	0.85	+0.73	+0.74
Owl	1.69	+0.43	+0.40
Pigeon	3.92	+0.06	−2.0

[a] Absolute visual threshold data from the following sources: man (Pirenne et al. 1957), cat (Gunter 1951), owl (Martin 1977), pigeon (Blough 1955). From Martin (1982)

For example, the maximal retinal illumination produced in the eyes of cat, owl, pigeon and rate are compared with that of man in Table 2. Table 3 compares the retinal image illumination (extended light source) and the mean absolute visual threshold in cat, owl and pigeon, expressed relative to those of man. Analyses of these led Martin (1982) to suggest that the retina of the owl (*Strix aluco*, a species considered to be highly nocturnal, Southern 1970), cat and man have attained the absolute limit of visual sensitivity in the vertebrate eye, at least for terrestrial species in which some kind of pattern vision is required at low light levels. However, the pigeon, despite its comparable retinal image brightness, is considerably less sensitive than cat, man or owl (Table 3). Analysis of parameters of typical rod photoreceptors in pigeon, owl and man (Table 4) leads to the conclusion that this difference in absolute visual threshold between the pigeon (considered a diurnal species) and the owl (considered a nocturnal species) is attributable to differences in retinal neural mechanisms, rather than to a property of individual retinal photoreceptors or different optical properties of the eyes in the two species (Martin 1982).

Thus, in summary, the large variation in naturally occurring luminance levels can be used to give a context to the relatively small differences in retinal image brightness revealed by comparison of *f*-numbers. These latter can also be used to elucidate something about similarities and differences in retinal mechanisms in the same species.

It is of interest to note the small difference in maximum retinal image brightness between vertebrate eyes often considered to represent the extremes of the diurnal-nocturnal dichotomy. In the case of pigeon and rat, this difference amounts to less than $\times 6$. Hughes (1977a, Fig. 10) has noted that diurnal and nocturnal eyes appear to differ at most by only fivefold in their maximum retinal illuminations, and in a direct comparison between the eyes of assumed nocturnal and diurnal forms of lizard, Citron and Pinto (1973) demonstrated a difference in image brightness of only twofold. This is a very small variation compared with the total range in luminance levels experienced between and within daytime and night-time

Table 4. Visual pigments, outer segment dimensions and mean absorbance of the rod photoreceptors of the owl, pigeon and man[a]

Species	Visual pigment λ_{max} (nm)	Outer segment (μm)		Specific absorbance of visual pigment (μm^{-1}) (k)	Mean absorbance (%)
		Mean length (x)	Diameter		
Owl	503	30	1.5	0.017	69
Pigeon	503	30	3–4	0.018	71
Man	498	25	1.5	0.019	66

[a] Visual pigment and outer segment dimension data are from the following sources: owl (Bowmaker and Martin 1978), pigeon (Bowmaker 1977), man (Bowmaker and Dartnall 1980). Mean absorbance (proportion of light incident on the receptor which is absorbed) calculated from the formula, $p = 1 - 10^{-kx}$ (axial illumination of outer segment is assumed). From Martin (1982)

(Fig. 8). This suggests that the basic simple design of the vertebrate eye is particularly efficient with respect to the brightness of the image which it produces and that the optical characteristics of the cornea and lens combinations are usually such as to maximize image brightness within a narrow range, irrespective of the absolute size of the eye.

4.1.2 Eye Size, Resolving Power and Information Capacity

Since the *f*-number is a ratio, it takes no account of the absolute size of the eye. However, the latter is quite variable (Table 2 and Fig. 6). The importance of focal length in relation to the separation of retinal photoreceptors in determining the resolving power of the eye has been well established. It has been considered in detail for a number of species, as also has the diffraction limit on resolution imposed by the absolute size of the pupil. These have been reviewed and discussed a number of times (e.g. Westheimer 1972; Muntz 1974; Miller 1979; Snyder 1979; Land 1981), and it would be redundant to rehearse them here.

It is perhaps worth considering briefly an example illustrating the need for caution in assuming visual capacity from focal length data, and the need to consider the totality of visual performance in order to understand the principles underlying optical design in a specific species.

It might be expected that the maximum resolving power of the owl eye would be superior to that of the pigeon, due to its greater focal length (Table 2 and Fig. 6). Minimum separable acuity data (Fig. 9) show that at photopic luminances the resolving power of the owl and pigeon eye are, in fact, very similar. Likewise, on the basis of comparable focal lengths, it might be expected that the eyes of man and the owl would be similar in their resolution. Figure 9 indicates that this, however, is not the case, with resolution in man $\times 5$ greater than in the owl. This suggests that, as is the cat's (Miller 1979), the owl's retinal image is undersampled by the photoreceptor array, and the maximum potential resolving power of the eye's optics is not realized.

This acuity data (Fig. 9) has been interpreted (Martin 1982) as suggesting that the owl's eye should not be viewed as simply nocturnal in character but designed like that of man, to function arrhythmically, i.e. over the full range of luminances, from photopic levels to the minimum absolute threshold. On the other hand, despite the similarity in image brightness and sensitivity of individual rod receptors to those of the owl (Sect. 4.1.1), the pigeon's is a diurnal eye, designed for optimal resolution at the limit dictated by its focal length only at relatively high luminance levels.

A more complete understanding of the principal optical characteristics of the vertebrate eye may require the ideas of the information capacity of eyes proposed by Snyder et al. (1977). They have shown that to optimize the amount of information extracted from the retinal image a different photoreceptor spacing is required at each mean luminance level and stimulus contrast. To achieve such optimal performance, individual photoreceptor responses may be neurally summed to provide 'effective photoreceptors' with a range of diameters. If a system of parallel processing is employed, then effective photoreceptors of varying diameters can be achieved within a single array of uniform-sized photoreceptors.

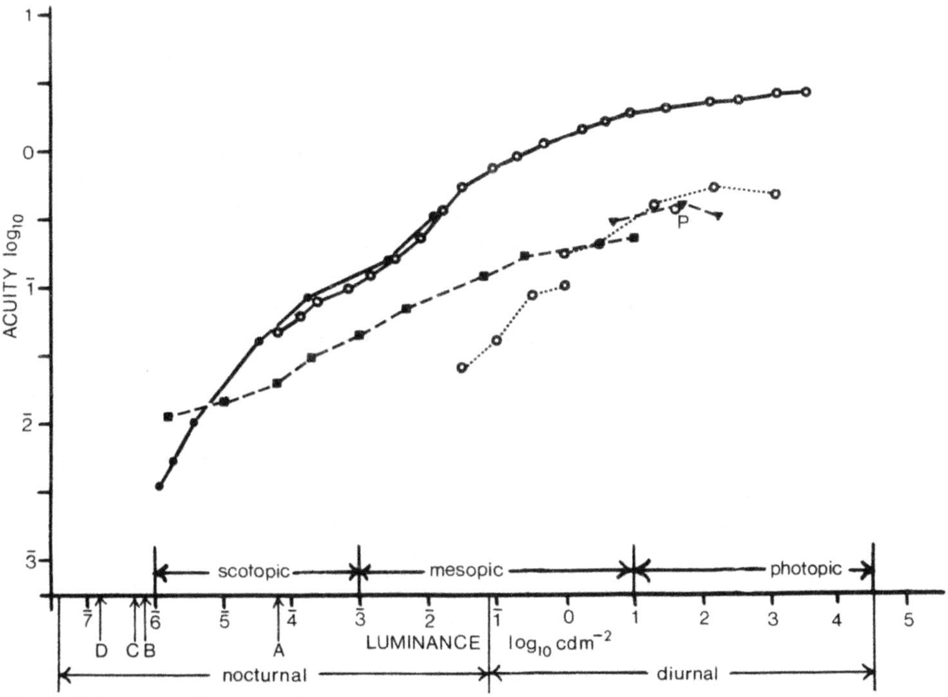

Fig. 9. Behaviourally determined minimum separable acuity data for man, pigeon and two owl species, *Strix aluco* and *Bubo virginianus.* Acuity is expressed as the $\log_{10}$ of the reciprocal of the minimum separable angle. The human data are for a Landholt C test stimulus of high contrast. Pigeon and owl data are for square wave high contrasts gratings. Photopic, mesopic and scotopic luminance ranges are indicated (Pirenne 1962, p 67); *A, B, C* and *D*: mean behaviourally determined absolute visual thresholds of pigeon, man, owl and cat, respectively (see Table 3). ●——●, man (Pirenne et al. 1957); ○——○, man (Shlaer 1937); ■– – –■, owl (Bubo, Fite 1973); ▼– – –▼, owl (Strix, Martin and Gordon 1974); ○*P*, pigeon (Columba, Blough 1971); ○------○, pigeon (Hodos and Leibowitz 1977, Hodos et al. 1971). From Martin (1982)

The range of effective photoreceptor diameters required for optimal information capacity, and their rate of increase in size with decreased luminance and contrast, is much greater at nocturnal than at diurnal luminance levels. For example, to cope optimally with the diurnal range of naturally occurring luminances presented in Fig. 8 (and a range of mean object contrasts between 1 and 0.003), Snyder et al. (1977, Fig. 10) predict the optimum range of photoreceptor spacing for the human eye to be from 0.008 to 0.2 degrees, a range of × 25. However, for the nocturnal luminance range which is of equal magnitude to the diurnal range ($\simeq$6.0 $\log_{10}$ units), the optimal photoreceptors diameters should vary between approximately 0.02 and 80 degrees, a range of × 4000. Thus, to function optimally over luminance ranges of comparable size but of different absolute value, the neural mechanisms for retinal image sampling in pigeon and owl require quite different degrees and flexibility of receptor integration.

Clearly, the potential for flexibility in the angular diameter of effective photoreceptors will increase as the retinal image is distributed over an increasing number of individual photoreceptors. Since there is an upper limit to photoreceptor packing density (Miller 1979), the only way the latter can be achieved is by the increase of retinal image size.

In eyes of different PND, but of equal photoreceptor packing density, the number of photoreceptors which subtend a given solid angle of visual space increases as the square of the ratio of the PNDs. Thus, assuming equal photoreceptor densities, the number of receptors per visual angle in (for example) the owl, *S. aluco*, is $\times 4.75$ greater than those subtending the same angle in the pigeon eye. Whether this theoretically greater potential for flexibility of image sampling in the owl is realized requires empirical investigation.

In the light of the above arguments, it has been hypothesized (Martin 1982) that the owl and human eyes, by virtue of their known acuity capacities (Fig. 9), are most suitably viewed as arrhythmic in function. The primary feature which enables these eyes to function in the nocturnal luminance range as well as the diurnal range is their long focal length which is a requisite for flexibility of neural sampling of the retinal image. Light-gathering power is thus viewed as being maximized only as a secondary feature of the eye (see Sect. 4.2).

4.1.2.1 Luminance Range as a Design Constraint. From the above discussion, it may be hypothesized that the primary constraint upon the vertebrate eye is the luminance *range* in which the eye is designed to function. Only at high light levels can high resolution be achieved (Snyder et al. 1977), and since it is well established that there is a finite limit to photoreceptor size and packing density (Miller 1979), high resolution can only be achieved with an absolutely long focal length. But equally, because of the need for flexibility of neural sampling of the retinal image, it can be argued that a long focal length will also be required by an eye whose principal function is to mediate visually guided behaviour over a wide luminance range, especially throughout the range of nocturnal luminance levels. Only a truly obligate nocturnal species can perhaps function adequately with an eye of very short focal length. Thus it has been suggested (Martin 1982) that the rat eye (anterior focal length = 3.33 mm) should be regarded as one of an obligate nocturnal species, designed to function optimally only within the lowest naturally occurring luminance and contrast levels which are experienced, for example, within forest floor habitats at night.

Regardless of the neural mechanisms used to sample the retinal image, it is not possible to be sensitive to high spatial frequencies (even at high contrast) at such low luminance levels (Snyder 1979). Therefore, for such limited stimulus conditions, only effective receptors of relatively large diameter are of adaptive value. [Minimum separable acuity measures for the rat range between 20 (Hermann 1958) and 52 (Lashley 1930) min of arc ($\bar{2}.7$ and $\bar{2}.3$ $\log_{10}$ acuity, respectively). Such acuity is not found in either man or the owl until well within the nocturnal luminance range (Fig. 9).] Such a relatively restricted range of large, angular diameter effective receptors can be readily achieved in a small eye by the neural summation of relatively few rod receptors. Thus, because of the particularly restricted luminance range of its visual environment, a small eye could be sufficient for optimal information capacity in the rat.

Corroborative support for the above suggestion can be found in the nocturnal bats, where Suthers and Wallis (1970) report eyes of smaller dimensions than that of the rat. The anterior focal lengths of *Myotis sodalis, Carollia perspicillata* and *Phyllostomus hastatus* are 1.04, 1.58 and 2.43 mm, respectively. A larger eye in these species would perhaps be a non-optimal solution to the problem of extracting the maximum amount of available spatial information from the animal's restricted natural visual environment. Thus, rather than regard the eyes of these bats as 'near vestigial' (Lythgoe 1979) because their visual capacities are much reduced compared with diurnal forms, it may be more enlightening to regard the visual environment as restricted with respect to the spatial information which can be extracted from it, and regard the eyes as optimally adapted to extract what little information is actually available.

A general schema of vertebrate eye design using the above arguments can be proposed thus: (a) Light level dictates the maximum amount of spatial information which can be extracted from a given visual scene, and (b) there exists a finite limit on photoreceptor size and packing density. Four general predictions follow:

1. Optimal spatial resolution at high light levels necessitates a long focal length.
2. Optimal spatial resolution over a wide range of light levels necessitates a long focal length.
3. Optimal spatial resolution at low light levels can be achieved with a short focal length.
4. Adequate, but not optimal, spatial resolution can be achieved at high light levels with a short focal length, but operation is restricted to high light levels.

From this the following predictions regarding the function of specific types of eyes can be made: Large-eyed animals are either strongly diurnal with resolution close to the theoretical maximum over a wide field of view (e.g. diurnal birds of prey), arrhythmic but with resolution below the theoretical maximum (e.g. owls, cats, ungulates), or arrhythmic with resolution close to the theoretical maximum attained only in a very localized part of the retina (e.g. man). Small-eyed animals are either strongly diurnal (e.g. the majority of passerine birds) or strongly nocturnal (e.g. rat).

None of these predictions are robust, and a specific problem is that there is more than one reason to predict the possession of both large and small eyes. This problem is not new, however. Walls (1942) (and many subsequent authors) gave two arguments which lead to the conclusion that eye size should be large. Thus by juxtaposing the visual capacities of resolution and sensitivity, it can be argued (a) that for high sensitivity without resolution loss, receptor size should be increased along with the focal length of the eye and (b) that for highest resolution, receptor size should be minimal while focal length is increased. Therefore, the problem is to distinguish amongst these different predictions by way of testable hypotheses or corroborative data.

Clearly, the arguments presented here can be clarified by the inclusion of schematic eye data for more eyes. However, what is also required are data on the visual performance and retinal structure of those eyes, together with analysis of how the eyes are actually used by the animal in its natural environment, e.g. the periods of activity of the animal in relation to naturally occurring luminance levels. This somewhat holistic approach to the problem of understanding the diversity of vertebrate eyes is a particularly laborious one, but one which may prove more fruitful than approaches based upon purely anatomical evidence.

4.1.3 Depth of Focus and Depth of Field

Knowledge of focal length and pupil diameter can be used to estimate the depth of focus and depth of field of the eye. These two parameters of an eye's performance can be used to make predictions about the necessity and amplitude of accommodatory mechanisms in an eye. If a sufficient amount is known about acuity as a function of luminance, then the depth of field at different luminance levels can also be predicted. These parameters can only be discussed in principle, since there do not appear to be any examples where specific predictions have been made and compared with actual performance of an eye. Campbell (1957) measured the depth of field of the human eye but specifically eschewed any theoretical analysis in terms of optical parameters because of its complexity. Hughes (1977b) discussed the depth of field of the rat eye in general terms in the context of the refractive state of that eye. Hughes (1977a) and Land (1981) have discussed the depth of focus in general terms, and Green et al. (1980) have developed formulae for calculating the approximate depth of focus of an eye from knowledge of its axial length.

Formulae describing in general terms the depth of focus and depth of field of an optical system are well known and are discussed in most general texts on geometrical optics. This is because of their particular practical importance in photographic and other imaging systems. The depth of focus and the depth of field are conjugate distances along the optic axis of the system in image and object space, respectively. The depth of focus is the permissible movement from an ideal image plane which can be made without causing a serious deterioration of the image. The problem arises in defining the criterion for a serious deterioration of the image, since this is a measure of the permissible blur of an image before it can be detected as out of focus. This is, of course, an empirical matter, and it is likely to be so small that it is not really in the realm of the assumptions of geometrical optics.

If the diameter of the permissible blur circle is z, then depth of focus (x') is given by the expression $x' = 2zf/A = 2z \cdot F_{no}$. The depth of field is the distance in object space equivalent to the depth of focus. Related to this is the hyperfocal distance (x) which is the distance from the optical system to the nearest object whose image is considered to be in focus on the retina. In the case of an emmetro-

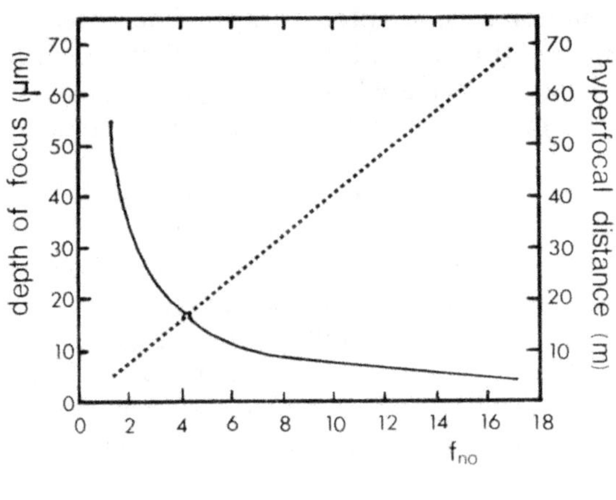

Fig. 10. Depth of focus (*dashed line* referring to *left-hand ordinate*) and hyperfocal distance (*solid line* referring to *right-hand ordinate*) as a function of *f*-number. Calculated from the formulae in the text, with the values $z = 2$ μm; $f = 17.0$ mm

pic eye, all objects from this point to infinity are in focus, and the hyperfocal distance can therefore be used to indicate for a given eye how close an object can be before accommodation becomes necessary. In the case of the emmetropic eye, the hyperfocal distance is approximately given by the expression $x = fA/2z$. Thus, as either focal length or aperture increases, the position of the nearest object considered to be in focus recedes from the eye. However, an increase in z brings that object point towards the eye.

Figure 10 shows the typical sort of relationship amongst depth of focus, hyperfocal distance and f-number in an emmetropic eye. The figure shows clearly that as f-number increases (pupil aperture becomes smaller), the hyperfocal distance decreases, i.e. more of the scene can be considered to be in focus. However, unless pupil aperture is made extremely small, some kind of accommodatory mechanism will be necessary in even small-eyed animals if they wish to receive focused images of objects close to them. It is also of interest to note that as acuity decreases (which can be considered equivalent to an increase in z) so does the need for accommodation. Thus as ambient illumination levels fall and the eye's resolving power decreases (Fig. 9), the hyperfocal distance may decrease considerably in an eye of fixed f_{no}.

4.1.4 Visual Fields

Schematic eye models have been used to estimate the limits of the optical visual field of the eye (Vakkur and Bishop 1963; Hughes 1972, 1979). The typical construction is shown in Fig. 11. The technique can not give more than a general indication of the limits of the field. Thus for a small pupil, Hughes (1979a) describes the limits of the rat uniocular field as lying between 147° (the minimum possible field) and 238° (the maximum possible field). In fact, the mean of these two measures (192.5°) is close to the observed extent of the uniocular field in this species (Hughes 1979a), but there is no theoretical or a priori reason to believe that this relationship will apply in other species. In such estimates the schematic eye is being used outside the limit of its paraxial assumptions, and this is probably part of the reason why the actual limits to the visual field have not been predicted with high accuracy. However, it should be noted that a general property of

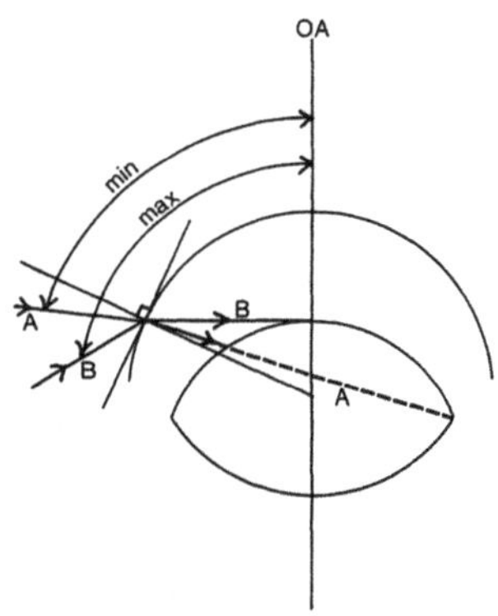

Fig. 11. Typical construction based upon schematic eye data used to estimate the limit of the optical field of an eye. It is assumed that the iris is not an obstruction. Ray *A* defines the minimum field; it enters the aqueous humour and is refracted in the direction of the opposite lens equator. A ray which adopts this path in the aqueous is highly likely to be refracted by the lens into the vitreous humour. Ray *B* in the aqueous humour grazes the lens at its anterior pole and hence cannot be refracted into the lens. Thus the actual limit to the optical field must lie between these two extremes. Symmetry of the visual field about the optic axis (*OA*) is assumed in estimating the total extent of the uniocular field

all optical systems is that visual field width is inversely related to focal length but in a complex manner. Visual field should therefore be considered a possible factor which has influenced eye design.

4.1.5 Focal Length and Eye Stability

While the maximum spatial resolution of an eye is determined primarily by its focal length and the characteristics of the photoreceptor array, maximal performance can only be achieved if the eye can fixate the object (which may be stationary or moving) with sufficient stability (Ludvigh 1947, 1948; Srinivasan and Bernard 1975). Stability is, of course, an ill-defined term but, as is well known from photography, the degree of camera movement permissible before deterioration of spatial resolution ensues is directly related to the focal length of the optical system and to the speed and resolution of the film. Thus the limits of performance of the oculomotor system (and the photoreceptor integration time) may also place a limit on the focal length of the eye in any particular species. A large eye functioning close to the theoretical limits of resolution will require high stability and fine muscular control, and this may be limited by the large globe's inertia.

It would seem reasonable to suppose that as eyes become larger the need for a specialized fixation device within the visual system may increase. The function of the deep convexiclivate fovea is enigmatic (Walls 1942; Pumphrey 1948; Kreithen and Keeton 1974; Snyder and Miller 1978; Harkness and Bennett-Clark 1978). Pumphrey postulated that the fovea functions to maintain accurate fixation or detect small angular movements of a fixed object. Comparative data (e.g. Meyer 1971; Miller 1979) would suggest that the most highly developed convexiclivate foveae also occur in the eyes with the longest focal lengths. If the need for increased stability in large eyes is accepted, then this could be interpreted as evidence in support of Pumphrey's hypothesis.

The above argument once again emphasizes the need for further comparative schematic eye data, which in this case may perhaps be usefully correlated with data on the occurrence and form of foveae. It may go some way in explaining the marked intra-retinal differences in foveal shape which are found in some avian species (Meyer 1977). For example, the eagle, *Aquila chrysaetos*, has both a deep, central convexiclivate fovea and a shallow, laterally placed one. It may be speculated that the former could serve primarily as the fixation device in this large eye, while the latter is simply regarded as an area of increased acuity in a behaviourally relevant part of the visual field.

4.2 The Variation of Ocular Design

Compared with the invertebrates (Land 1981), the variation in optical design of the vertebrate eye may appear on first analysis to be rather uninteresting. All of the schematic eyes in Fig. 6 show the same basic structure. However, as Fig. 1 indicates, there is considerable variation within this one design. It would seem unlikely that these eyes are the result of a random process, and their structure probably reveals some optimal solution to optical design problems, given the anatomical constraints of the two optical systems (cornea and lens) which are involved.

Explanatory generalizations of the diversity of vertebrate ocular designs as adaptations to different visual environments have frequently employed the nocturnal-diurnal dichotomy. Thus, Walls (1942), Rochon-Duvigneaud (1943), Duke-Elder (1958), Tansley (1965) and Hughes (1977) have all offered interpretations of eye design within such a framework. The majority of these interpretations have been based upon assumed or inferred optical properties and visual capacities, rather than upon schematic eye and psychophysical data. However, as argued in the above section, such a rigid dichotomy in the function and design of eyes, and in the properties of the visual environment, may be too simple. Indeed, Lythgoe (1979) has described in detail many aspects of the natural visual environment, some or all of which could influence eye design.

In interpreting the variation of ocular design, it is worth recalling that optical systems of the same fundamental parameter, focal length, can be achieved by an almost infinite combination of optical elements. It is this fact which makes optical design an art as much as a science, with the final solution to a particular design process being arrived at by a series of successive approximations and empirical evaluations, rather than a straightforward application of formulae (Brown 1965). It is also this fact which poses special problems for the visual scientist interested in attempting to discern 'rules' of optical design in vertebrate eyes. Thus, the need in such an exercise for a wide comparative and quantitative context is obvious.

Some insight into the possible design constraints upon vertebrate eyes, which may exist independently of any general functional considerations such as the nocturnal-diurnal division, can be gained from the following analysis. This is based initially upon simple quantitative measures of eye design (from schematic models), rather than upon assumed properties, and depends to some extent upon the discursive technique of reductio ad absurdum. No attempt is made to determine an optimal solution to these design problems. It is hoped that sufficient schematic and anatomical data will eventually be accumulated to allow the quantification of the various factors considered. This will eventually permit, by way of mathematical modelling, a more rigorous test of the various hypotheses.

4.2.1 Focal Length and Lens: Cornea Power

Between the owl and the rat (Fig. 6), there is a fivefold variation in refractive power (or focal length) and a 4.5-fold variation in axial length. Also between these two species, there is a marked difference in the relative contributions of the refractive powers of the lens (F_L) and cornea (F_C) to the total power of the system (Fig. 12). Thus in the rat, lens power : cornea power ($F_L : F_C$) is 2.16, while in the owl this ratio is .83. For the seven species in which lens and corneal powers have been schematically and independently determined, an approximately linear relationship seems to exist between the focal length of the eye and the ratio of the optical powers of lens to cornea (Fig. 12). Such a relationship may possibly indicate the operation of a common design constraint amongst these eyes. However, within a particular focal length, some variation of lens : cornea power may

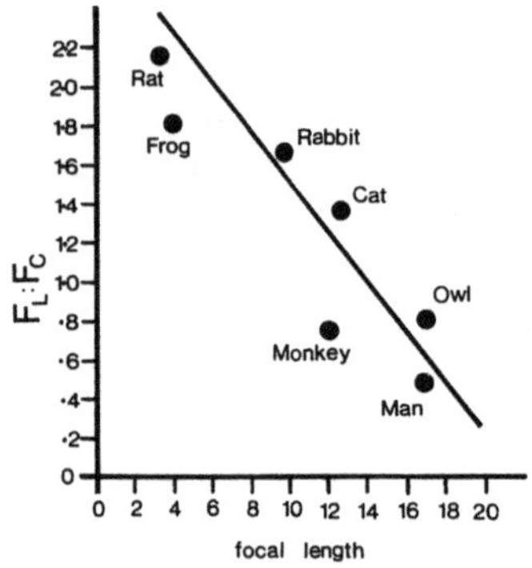

Fig. 12. The ratio of lens: cornea refractive powers (F_L: F_C) as a function of anterior focal length in the schematic eyes of seven species for which F_L and F_C have been determined independently. The *straight line* has been fitted by eye. Data from the sources given in Table 1

also be found, and this is exemplified by the comparison of the eyes of man and an owl. Given the range of intersubject variability (Sect. 3.1.4), these two eyes may be considered to have identical refractive powers, yet the F_L: F_C ratios are .47 and .83 in man and owl, respectively.

Some of the problems of discerning the constraints which may underly the relationship of Fig. 12 and of generally discerning the design rules of the vertebrate eye are indicated in Figs. 13, 14 and 15.

4.2.2 Possible Variations in Lens Design

Figure 13 shows nine different schematic lenses, all of which have the same calculated refractive power (50 D). The lenses are of homogeneous refractive index, and three values comparable to those used in schematic eye models are shown. The lenses are biconvex and symmetrical with surface powers of either 26, 27 or 28 D. The lens parameters have been calculated using the simple formulae given in the caption. Clearly, some of these lenses are absurd solutions to the problem of designing a 50 D lens, but it is also clear that a number represent quite acceptable solutions. However, each lens differs in its thickness and equatorial diameter and would also show variation in its image quality. Equatorial diameter would in turn also determine the relative brightness of the image so produced. Since each lens has the same focal length, relative image brightness will be proportional to the square of their apertures (aperture in this case approximated by equatorial diameter). Thus in these lenses image brightness would vary over a range of about $\times$ 12 (1.1 $\log_{10}$).

Given that these are just nine sample lenses selected from a continuum, it is clear that the problem of optimal design is complicated. If we assume that when designing an eye the choice is made that a 50 D lens is required, a second criterion to aid the choice of a particular type from the array of possible lenses in Fig. 13 might be that light-gathering power should be maximal. However for a given focal length, an increase in image brightness is only obtained by increasing lens size (both equatorial diameter and thickness), and this is likely to be restricted in the case of the eye by anatomical and physiological considerations, as well as straightforward mechanical ones.

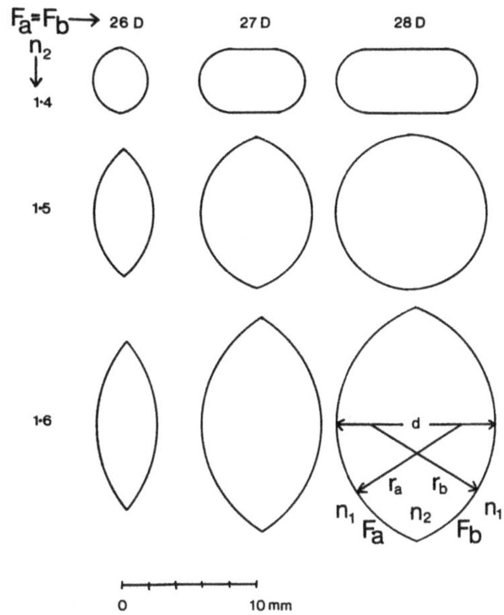

Fig. 13. A range of schematic lenses, all of which have the same refractive power (F) of 50 D when immersed in a medium of refractive index (n_1) which is similar to that of aqueous and vitreous humour. The lenses are symmetrical, i.e. the two surfaces have the same curvatures ($r_a = r_b$) and hence the same refractive powers ($F_a = F_b$). Each column shows three lenses of the same surface power, and each row shows three lenses of the same refractive index (n_2). The three values of n_2 cover the range of assumed homogeneous refractive index used in schematic eye models. The characteristics of these lenses have been calculated from the standard formulae thus: $F_a = (n_2 - n_1)\,1000/r_a$; $F_b = (n_1 - n_2)\,100/-r_b$ and $F = F_a + F_b - d(10^{-3})F_a \cdot F_b/n_2$ [d (lens thickness) and r in mm]. Since the focal lengths of these lenses are identical, it is clear that there will be differences in image brightness amongst them. Relative image brightness will be approximately proportional to the ratio of the square of the lens equatorial diameters, thus the difference in image brightness between the largest and smallest lens shown here will be about 12-fold. In addition, the lenses will vary in their image quality

It should be borne in mind that these examples are based upon symmetrical lenses of homogeneous refractive index, and although this is of considerable computational convenience, it is far from the case in real vertebrate lenses. Since it is not known in detail how the refractive index gradient of the vertebrate lens is achieved anatomically, how it might be related to lens size, or how the gradient is related to the assumed homogeneous bulk index, these arguments can be considered as only illustrative of the problem. For example, it is not known how much variation may be possible in the refractive index gradient within a lens of particular dimensions. Interspecific variations in refractive index gradients exist (Sect. 3.1.2), but it is not known whether these are related in any way to the absolute dimensions of the lens.

4.2.3 Possible Variation in Eye Design

As indicated above, a particular lens of a given focal length may be selected for use in a specific application according to such criteria as its *f*-number and size. When this lens has to be coupled with another component, such as the cornea, other considerations may also come into play.

Figures 14 and 15 give some hypothetical eye designs for the rat and man. The designs are framed within a number of simple assumptions but represent eyes whose ratios of lens: cornea powers are reversed from the real case, i.e. the hypothetical rat eyes have an $F_L : F_C$ ratio of .46 (approximately that found in the real

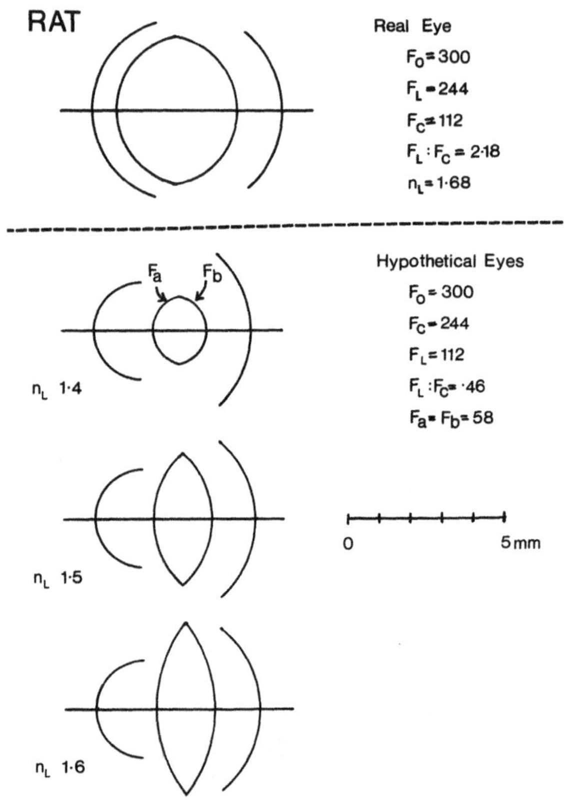

Fig. 14. The real schematic eye of the rat and three hypothetical schematic rat eyes. Each hypothetical eye has the same total power (F_O) as the real eye, but the powers of the cornea (F_C) and lens (F_L) have been reversed from their real values. (All powers are in dioptres.) Thus in the real rat eye, $F_L : F_C = 2.18$; in the hypothetical eyes, $F_L : F_C = 0.46$ (this value is approximately the same as that in the real human eye, Fig. 15). The hypothetical cornea is a single surface of the same refractive index as the real cornea. The hypothetical lenses are symmetrical with surface powers ($F_a = F_b$) of 58 D in each eye. The only parameter which changes between the hypothetical eyes is the bulk refractive index (n_L) of the lens. It is clear how changes in this one parameter result in eyes of quite different design. Reversing the lens: cornea ratio from that of the real eye results in a marked change in the structure of the schematic eye, yet the fundamental parameter (*f*) of these four schematic eyes is the same

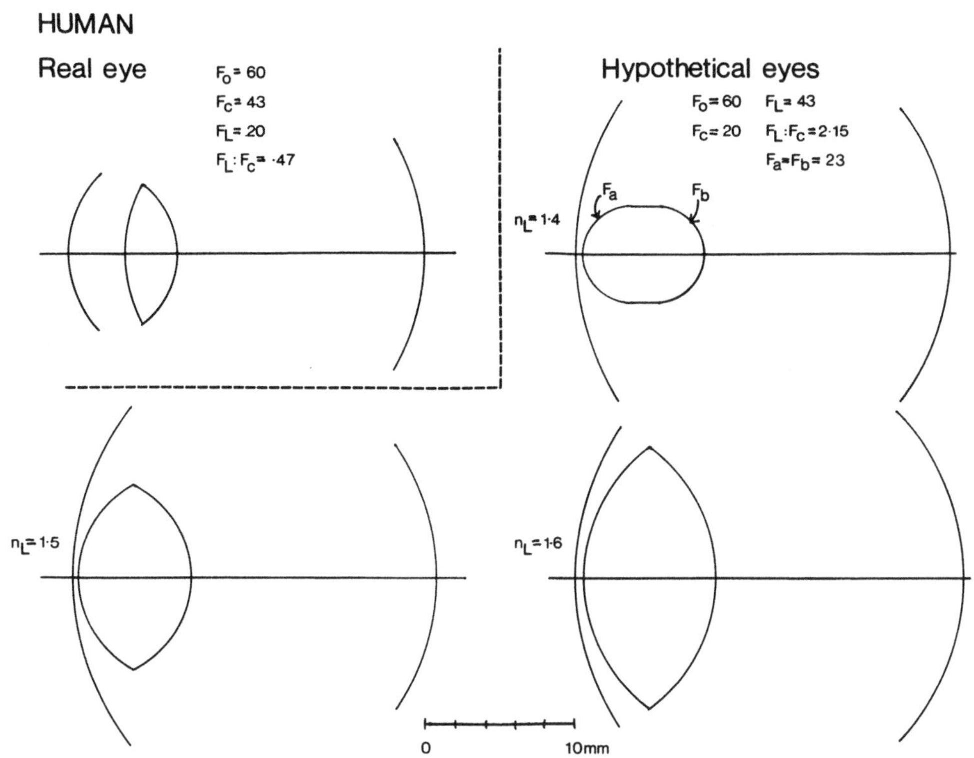

Fig. 15. The real schematic human eye and three hypothetical schematic human eyes. The same procedure and assumptions have been followed in constructing the hypothetical human eyes as are described for the case of the rat in Fig. 14. It is of interest to note that the hypothetical $F_L : F_C = 2.15$, approximately the same as the ratio in the real rat eye

human eye), while the human eye has a ratio of 2.15 (approximately the same as that found in the real rat eye). The focal lengths of the hypothetical eyes are the same as the real eyes. The cornea: lens ratio fixes the power of the cornea and hence, for a particular refractive index, fixes the corneal radius of curvature (the cornea is assumed to be a single surface of refractive index 1.376). Three different lenses, all with the same surface powers but with different bulk refractive indices, are shown for each eye. Image position, and hence the axial length of the eye, change only by an amount equal to the change in lens thickness between the models.

In the case of the rat eye (Fig. 14), the increased curvature of the cornea brings the principal points of the system forward, compared with the real eye, and the axial length is shorter. The anterior chamber is much deeper than in the real eye, and this may present problems of anatomical structure and mechanical stability. The most salient feature of these model eyes compared with the real rat eye is the reduced maximum aperture size in the models, and this is dictated by the corneal curvature. While it would be possible to adopt a lens refractive index value which would make the lens and cornea of comparable diameter, and hence the structure could be made more anatomically stable and optically efficient in terms of image brightness (i.e. all light captured by the cornea passes into the lens), that brightness could not equal that found in the real eye.

Figure 15, the human eye models, illustrates changes in optical parameters which are, in essence, the reverse of the changes described for the rat eye models. However, compared with the real eye, these models would also seem to be inferior solutions to the optical design problem of an eye with the same focal length as the human eye. For example, anterior chamber depth is reduced and the corneal curvature decreased, and this may present anatomical and mechanical problems. While maximum image brightness is not necessarily reduced by the change in corneal curvature (it could, in fact, be increased), a considerably larger lens is required.

4.2.4 Focal Length and Other Design Constraints

These two examples of alternative eye models seek to illustrate some of the design considerations which may have influenced the evolution of presently extant vertebrate eye forms. In particular, they illustrate that the relative contributions of lens and cornea may be primarily a function of the focal length of the eye, with an increasing proportion of the eye's refractive power being produced by the lens as focal length decreases. However, the actual eye structure seen is probably a complex result of anatomical, physiological and optical factors, the salience of which cannot be determined at present, but which may be eventually elucidated by the accumulation of schematic eyes for a wider diversity of species. Possibly the most important optical, as opposed to anatomical, factor has been that for a given focal length image brightness should be maximized. It has been argued (see, for example, Srinivasan and Bernard 1975; Snyder 1979; Land 1982) that in any eye, no matter what the visual task, there is always advantage in having the brightest possible retinal image, except at the very highest levels. This would lead to the prediction that a large variation in maximum image brightness is unlikely to be found amongst the vertebrates (see Sect. 4.1.1). A second optical factor is the way in which the changes illustrated in Figs. 14 and 15 may alter the visual field of the eye (Sect. 4.1.4). There may also be limitations regarding the degree of accommodation that is necessary in an eye of a given power and *f*-number (Sect. 4.1.3) and what accommodative changes are possible by (say) deformation of the lens of a certain size and shape.

Moderating these optical considerations may be mechanical, anatomical and physiological factors which impose limits on both the relative and absolute sizes or corneae and lenses and also on the absolute size of the eye itself. Given the differences in eye structure amongst the vertebrate classes [see, for example, Walls (1942); Duke-Elder (1958); Meyer (1977)], these latter factors may operate within different limits in each class. These factors may also pose a limit as to what can be achieved by way of optical adaptations to the problems imposed by vision in different types of visual environment, especially in amphibious species (see Sivak 1978, for a review).

All of these arguments suppose that absolute size of the eye by way of focal length is of prime importance in determining the basic optical design of the eye. However, as suggested above, alternative approaches are possible. Thus Hughes

(1977a) has recently extended the reasoning of Walls (1942) and argued from an analysis of four scaled, schematic eyes that in mammals the relative contributions of lens and corneal powers are, in fact, independent of the absolute size of the eye but are related to a nocturnal-diurnal division in function (but see Sect. 4.1.1). However, even if this functional dichotomy in optical design is accepted, then it might be argued that the series owl, cat, rabbit and rat should all fall within the former functional category, and hence on the basis of this hypothesis, more uniformity in the $F_L:F_C$ ratio would be expected amongst these species (Fig. 12).

In addition, two further, more specific, interpretations of the relationship presented in Fig. 12 are also possible. First, ignoring the arguments presented in Sect. 4.1, it could be proposed that the specific slope of the straight line in Fig. 12 presents a relationship found amongst eyes which are basically 'nocturnal' in design, and that a different relationship might be found if a similar analysis were conducted on 'diurnal' eyes of different focal length. Second, Fig. 12 could be considered as presenting primarily a relationship found amongst mammalian eyes, and a different relationship may be found amongst the eyes of species, from different vertebrate taxa.

Thus, whether the relationship shown in Fig. 12 actually presents a fundamental property of the terrestrial vertebrate optical system clearly requires schematic eye data for many more species. The inclusion of such data does hold out the possibility of more rigorous and testable hypotheses to account for at least some of the diversity of vertebrate eye structure.

5 Conclusion

The aim of this chapter has been to summarize the value of schematic eye models as a tool for understanding how the vertebrate eye works as an optical system and also as a means of providing some explanatory generalizations concerning eye structure in vertebrates. It has been argued that schematic eye models are founded on many assumptions, but from these models common principles of optical design can be discerned and, in some cases, hypotheses can be formulated which go some way to account for these principles.

Throughout, it has been emphasized that schematic eye models for more species are required in order to broaden the comparative context in which results can be evaluated. There is an especial need for data on species which are neither domesticated nor laboratory strains, and in particular, more birds, reptiles and amphibia should be investigated. In addition, there would seem to be an important need for a more thorough understanding of the nocturnal-diurnal distinction which has been central to hypotheses in visual science for over a century. The sensory, ecological and behavioural aspects of this simple and readily adopted classification of vertebrate species deserves close attention.

The recent development of models which can deal with non-paraxial optics is most welcome. So much of interest in the vertebrate eye seems to take place away from the optic axis, but this cannot be discussed here because of the limitations of the optical models. Nevertheless, the peripheral structures pose important

optical problems for the understanding of vertebrate eyes. There are, for example, marked asymmetries of eye structure about the optic axis in many species, and foveae and other specialized areas occur frequently in the retinal periphery (especially amongst avian species). Also, the majority of vertebrates are lateral eyed. This means that vision in the frontal visual field, which is used to control behaviour during rapid forward motion, is subserved by peripheral optics.
Regardless of the merits of the hypotheses discussed here, it is hoped that the reader will appreciate the fundamental value of the schematic eye in revealing something of the remarkable subtlety of optical design which is to be found within every vertebrate eye.

References

Blest AD, Land MF (1977) The physiological optics of *Dinopis subrufus*. Koch L: A fish-lens in a spider. Proc R Soc Lond B 196:197 – 222

Block MT (1969) A note on the refraction and image formation of the rat's eye. Vision Res 9:705 – 711

Blough DS (1955) Method for tracing dark adaptation in the pigeon. Science 121:703 – 704

Blough PM (1971) The visual acuity of the pigeon for distant targets. J Exp Anal Behav 15:57 – 68

Bobier CW, Sivak JG (1978) Chromoretinoscopy. Vision Res 18:247 – 250

Bonnet R (1964) La topographie cornéenne. Desroches, Paris

Bowmaker JK (1977) The visual pigments, oil droplets and spectral sensitivity of the pigeon. Vision Res 17:1129 – 1138

Bowmaker JK, Dartnall HJA (1980) Visual pigments of rods and cones in a human retina. J Physiol (Lond) 298:501 – 511

Bowmaker JK, Martin GR (1978) Visual pigments and colour vision in a nocturnal bird, *Strix aluco* (tawny owl). Vision Res 18:1125 – 1130

Brown EB (1965) Modern optics. Reinhold, New York

Campbell FW (1957) The depth of field of the human eye. Opt Acta 4:157 – 164

Campbell MCW, Hughes A (1981) An analytic, gradient index schematic lens and eye for the rat which predicts aberrations for finite pupils. Vision Res 21:1129 – 1148

Charman WN, Jennings JAM (1976) Objective measurements of the longitudinal chromatic aberration of the human eye. Vision Res 16:999 – 1005

Charman WN, Tucker J (1973) The optical system of the goldfish eye. Vision Res 13:1 – 8

Citron MC, Pinto LH (1973) Retinal image larger and more luminous for a nocturnal than for a diurnal lizard. Vision Res 13:873 – 876

Davson H (1962) Visual optics and the optical space sense. In: The eye, vol 4. Academic, New York

Duke-Elder S (1958) The eye in evolution. System of ophthalmology, vol 1. Henry Kimpton, London

DuPont J, De Groot PJ (1976) A schematic dioptric apparatus for the frog's eye. Vision Res 16:803 – 810

Emsley HH (1948) Visual optics, 4th edn. Hatton, London

Federer CA, Tanner CB (1966) Spectral distribution of light in the forest. Ecology 47:555 – 560

Fincham WHA, Freeman MH (1980) Optics, 9th edn. Butterworths, London

Fite KV (1973) Anatomical and behavioural correlates of visual acuity in the great horned owl. Vision Res 13:219 – 230

Fletcher A, Murphy T, Young A (1954) Solutions of two optical problems. Proc R Soc Lond A 223:216 – 225
Gauss JKF (1841) Dioptrische Untersuchungen. Göttingen
Glickstein M, Millodot M (1970) Retinoscopy and eye size. Science 168:605 – 606
Granda AM, Dvorak CA (1977) Vision in turtles. In: Crescitelli F (ed) Handbook of sensory physiology, vol VII/5. Springer, Berlin Heidelberg New York
Gullstrand A (1909) Appendix. In: Von Helmholtz H Handbuch der physiologischen Optik, 3rd edn.
Gunter R (1951) The absolute threshold for vision in the cat. J Physiol (Lond) 119:8 – 15
Harkness L, Bennet-Clark HC (1978) The deep fovea as a focus indicator. Nature 272:814 – 816
Hermann G (1958) Beiträge auf Physiologie des Rattenauges. Z Tierpsychol 15:463 – 517
Hirschberg J (1882) Zur Dioptrik und Ophthalmoskopie der Fisch- und Amphibienaugen. Arch Anat Physiol 6:493 – 526
Hodos W, Leibowitz RW (1977) Near-field visual acuity of pigeons: effects of scotopic adaptation and wavelength. Vision Res 17:463 – 467
Hodos W, Leibowitz RW, Bonbight JC (1976) Near-field acuity of pigeons: effects of head position and stimulus. J Exp Anal Behav 25:129 – 141
Howcroft MJ, Parker JA (1977) Aspheric curvature for the human lens. Vision Res 17:1217 – 1223
Hughes A (1972) A schematic eye for the rabbit. Vision Res 12:123 – 138
Hughes A (1977a) The topography of vision in mammals of contrasting life style: Comparative optics and retinal organisation. In: Handbook of Sensory Physiology, vol 7/5, Crescitelli F (ed). Springer, Berlin Heidelberg New York
Hughes A (1977b) The refractive state of the rat eye. Vision Res 17:927 – 939
Hughes A (1979a) A schematic eye for the rat. Vision Res 19:569 – 588
Hughes A (1979b) A useful table of reduced schematic eyes for vertebrates which includes computed longitudinal chromatic aberrations. Vision Res 19:1273 – 1275
Hughes A (1979c) The artefact of retinoscopy in the rat and rabbit eye has its origin at the retina/vitreous interface rather than in longitudinal chromatic aberration. Vision Res 19:1293 – 1294
Jansson F (1963) Measurements of intraocular distances by ultrasound. Acta Ophthalmol [Suppl] (Copenh) 74
Kirschfeld K (1974) The absolute sensitivity of lens and compound eyes. Z Naturforsch 29c:592 – 596
Kreithen ML, Keeton WT (1974) Detection of polarized light by the homing pigeon *Columba livia*. J Comp Physiol 89:83 – 92
Land MF (1981) Optics and vision in invertebrates. In: Handbook of sensory physiology, vol VII/6B, Autrum H (ed). Springer, Berlin Heidelberg New York
Lashley KS (1930) The mechanism of vision III. The comparative visual acuity of pigmented and albino rats. J Gen Physiol 37:481 – 484
Longhurst RS (1973) Geometrical and physical optics. Longman, London
Lopping B, Weale RA (1965) Changes in corneal curvature following ocular convergence. Vision Res 5:207 – 215
Lotmar W (1971) Theoretical eye model with aspherics. J Opt Soc Am 61:1522 – 1529
Lotmar W, Lotmar T (1974) Peripheral astigmatism in the human eye. Experimental data and theoretical model predictions. J Opt Soc Am 64:510 – 513
Ludvigh E (1947) Visibility of the deer fly in flight. Science 105:176 – 177
Ludvigh E (1948) The visibility of moving objects. Science 108:63 – 64
Lythgoe JN (1979) The ecology of vision. Clarendon, Oxford
Marshall J, Mellerio J, Palmer DA (1973) A schematic eye for the pigeon. Vision Res 13:2449 – 2453
Massof RW, Chang FW (1972) A revision of the rat schematic eye. Vision Res 12:793 – 796

Martin GR (1977) Absolute visual threshold and scotopic spectral sensitivity in the tawny owl, *Strix aluco*. Nature 268:636 – 638
Martin GR (1982) An owl's eye: schematic optics and visual performance in *Strix aluco* L. J Comp Physiol 145:341 – 349
Martin GR, Gordon IE (1974) Visual acuity in the tawny owl *(Strix aluco)*. Vision Res 14:1393 – 1397
Matthiessen L (1880) Untersuchungen über den Aplanatismus und die Periscopie der Kristallinsen in den Augen der Fische. Pfluegers Arch 21:287 – 307
Matthiessen L (1886) Ueber den physikalisch-optischen Bau des Auges der Cetaceen und der Fische. Pfluegers Arch 38:521 – 528
Meyer DB (1977) The avian eye and its adaptations. In: Crescitelli F (ed) Handbook of sensory physiology, vol VII/5. Springer, Berlin Heidelberg New York
Miller WH (1979) Ocular optical filtering. In: Autrum H (ed) Handbook of sensory physiology, vol VII/6A. Springer, Berlin Heidelberg New York
Millodot M (1972) Reflection from the fundus of the eye and its relevance to retinoscopy. Fondazione Giorgio Ronchi 27:31 – 50
Millodot M, Sivak J (1978) Hypermetropia of small animals and chromatic aberration. Vision Res 18:125 – 126
Muntz WRA (1974) Comparative aspects in behavioural studies of vertebrate vision. In: Davson H (ed) The eye, vol 6. Academic, New York
Nakao S, Fujimoto S, Nagata R, Iwata K (1968) Model of refractive index distribution in the rabbit crystalline lens. J Opt Soc Am 58:1125 – 1130
Nakao S, Mine K, Nishioka K, Kamiya S (1969) The distribution of refractive indices in the human crystalline lens. Jpn J Ophthalmol 23:41 – 44
Nakao S, Mine K, Nishioka K, Kamiya S (1970) A new schematic eye and its clinical application. Abstract of the 21st international congress of Ophthalmology, Mexico
Natural illumination charts (1952) US Navy research and development project NS 714-100 Report No. 374-1 (September)
Nuboer JFW, van Genderen-Takken H (1978) The artifact of retinoscopy. Vision Res 18:1091 – 1096
Philipson B (1969) Distribution of protein within the normal rat lens. Invest Ophthalmol Vis Sci 8:258 – 269
Pirenne MH, Marriott FHC, O'Doherty EF (1957) Individual differences in night vision efficiency. Med Res Counc (GB) Spec Rep Ser 294
Pomerantzeff O, Fish H, Govignon J, Schepens CL (1971) Wide-angle model of the human eye. Ann Ophthalmol 3:815 – 819
Pumphrey RJ (1948) The theory of the fovea. J Exp Biol 25:299 – 310
Pumphrey RJ (1961) Concerning vision. In: Ramsay JA, Wigglesworth VB (eds) The cell and the organism. Cambridge University Press, Cambridge
Rivamonte A (1977) The under-corrected lens of the frog eye *(Rana esculenta)* could yield comparable aerial and underwater vision. Vision Res 17:1237 – 1238
Rochon-Duvigneaud A (1943) Les yeux et la vision des vertébrés. Masson, Paris
Rushton RH (1938) The clinical measurement of the axial length of the living eye. Trans Ophthalmol Soc UK 58:136
Shlaer S (1937) The relation between visual acuity and illumination. J Gen Physiol 21:165 – 188
Sivak JG (1976a) Optics of the eye of the 'four-eyed fish' *Anableps anableps*. Vision Res 16:531 – 534
Sivak JG (1976b) The accommodative significance of the "ramp" retina of the eye of the sting-ray. Vision Res 16:945 – 950
Sivak JG (1976c) The role of the flat cornea in the amphibious behaviour of the blackfoot penguin. Can J Zool 54:1341 – 1345
Sivak JG (1977) The role of the spectacle in the visual optics of the snake eye. Vision Res 17:293 – 298

Sivak JG (1978) A survey of vertebrate strategies for vision in air and water. In: Ali MA (ed) Sensory ecology. Plenum, New York
Sivak JG, Allen DB (1975) An evaluation of the "ramp" retina of the horse eye. Vision Res 15:1353–1356
Sivak JG, Millodot M (1977) Optical performance of the penguin eye in air and water. J Comp Physiol 119:241–247
Sivak JG, Bobier WR, Levy B (1978) The refractive significance of the nictitating membrane of the bird eye. J Comp Physiol 125:335–339
Snyder AW (1979) Physics of vision in compound eyes. In: Autrum H (ed) Handbook of sensory physiology, vol VII/6A. Springer, Berlin Heidelberg New York
Snyder AW, Miller WH (1978) Telephoto lens system of falconiform eyes. Nature 275:127–129
Snyder AW, Laughlin SB, Stavenga DG (1977) Information capacity of eyes. Vision Res 17:1163–1175
Soemmering ST (1818) De oculorum hominis animaliumque sectione horizontali commentatio. Vanderhoeck and Ruprecht, Goettingen
Sorsby A, Benjamin B, Davey JB, Sheridan M, Tanner JM (1957) Emmetropia and its aberrations. Med Res Counc (GB) Spec Rep Ser 293
Sorsby A, Benjamin B, Sheridan M (1961) Refraction and its components during the growth of the eye from the age of three. Med Res Counc (GB) Spec Rep Ser 301
Southern HN (1970) The natural control of a population of tawny owls *(Strix aluco)*. J Zool 162:197–285
Srinivasan MV, Bernard GD (1975) The effect of motion on visual acuity of the compound eye: a theoretical analysis. Vision Res 15:515–525
Suthers RA, Wallis NE (1970) Optics of the eyes of echo-locating bats. Vision Res 10:1165–1173
Tansley K (1965) Vision in vertebrates. Chapman Hall, London
Taylor AH, Kerr GP (1941) Distribution of energy in daylight. J Opt Soc Am 31:3–8
Tscherning M (1898) Optique physiologique. Paris. English translation by Weiland C (1920) Physiological optics. Keystone, Philadelphia
Vakkur GJ, Bishop PO (1963) The schematic eye in the cat. Vision Res 3:357–381
von Helmholtz H (1909) Handbuch der physiologischen Optik, 3rd edn. Optical Society of America
Wallman J, Turkel J, Trachtman J (1978) Extreme myopia produced by modest change in early visual experience. Science 201:1249–1251
Walls GL (1942) The vertebrate eye and its adaptive radiation. Cranbrook Institute of Science, Michigan
Weekers R, Delmarcelle Y, Luyckx J (1975) Biometrics of the crystalline lens. In: Bellows JG (ed) Cataract and abnormalities of the lens. Grune and Stratton, New York
Westheimer G (1970) Image quality in the human eye. Opt Acta 17:641–658
Westheimer G (1972) Optical properties of vertebrate eyes. In: Fuortes MGF (ed) Handbook of sensory physiology, vol VII/2. Springer, Berlin Heidelberg New York
Young T (1801) On the mechanism of the eye. Philos Trans R Soc Lond [Biol] 92:23–88

Organization of the Outer Plexiform Layer of the Tetrapoda Retina*

A. Gallego

Departamento de Fisiología, Facultad de Medicina, Universidad Complutense, Madrid-3, Spain

* The preparation of the manuscript as well as much of the personal research involved was supported by grant n° 2796/76 of the Comisión Asesora de Investigación Científica y Técnica (Spain)

1 Introduction

The study of the functional organization of the first synapse of the centripetal visual pathway at the outer plexiform layer level (OPL) ought to be made through the application of combined histological, electrophysiological, and neurochemical techniques. A large amount of new evidence has been accumulated in the past 20 years on the structure of the retina and on the electrical responses of retinal cells to light stimulus. Also, recently, many substances considered as neurotransmitters in the brain have been found in the retina. The goal of the study of retinal function is to integrate the data obtained by structural and electrophysiological techniques and to identify and determine the role played by neurotransmitters or neuromodulators in the function of the retina.
In this study it is important to realize the morphological and biochemical diversity displayed by the visual cells in the vertebrate retina which, according to Crescitelli (1972), has been produced "through the interaction of natural selection with diversity in the photic environment." The evidence obtained shows that bipolar and especially horizontal cells, closely related to visual cells, display morphological and probably biochemical differences among classes, genus, and even species according to the photic environment. These differences give peculiarities to the organization of the OPL, which must be taken into account when studying a particular retina with electrophysiological or neurochemical techniques.

2 Early Studies

The first period of the study of the retina was dominated by the histological work and contributions of Cajal. He showed that the retina was formed by individual nerve cells, and he supplanted the long-dominant, nineteenth century concept of continuity of the optic nerve fibers through the retina up to the visual cells with a new, revolutionary concept of chains of nerve cells connceted by "gearing dispositives" (synapses). At the level of the OPL, his basic findings can be summarized in the description of the "feet" of the visual cells (synaptic bodies) and their location in the OPL; the discovery of cone and rod bipolar cells, giving support to Schultze's duplicity theory; and the discovery that the cells described previously as "basal," "stellate," "asteriform," "subreticular," or "concentrical" were actually short axon neurons. This cell type was named "horizontal cell" (HC) by Cajal (1893), who described two varieties, "external" and "internal" in mammals and amphibian retina, and "brush" and "stellate" in the avian and reptile retina.
Cajal summarized his studies on the structure of the retina, establishing in his schematic drawings two centripetal pathways; a centrifugal path for cones and rods; and two systems of lateral connections at the level of the two plexiform layers mediated by HC and amacrine cells.
Cajal's concepts remained unquestioned for 75 years, in spite of the problems he himself raised in a paper published a year before his death (1933). In this paper he discussed what he considered the "horizontal cell paradox" and the "enigma of the amacrine cells." Referring to horizontal cells, he had concluded that they

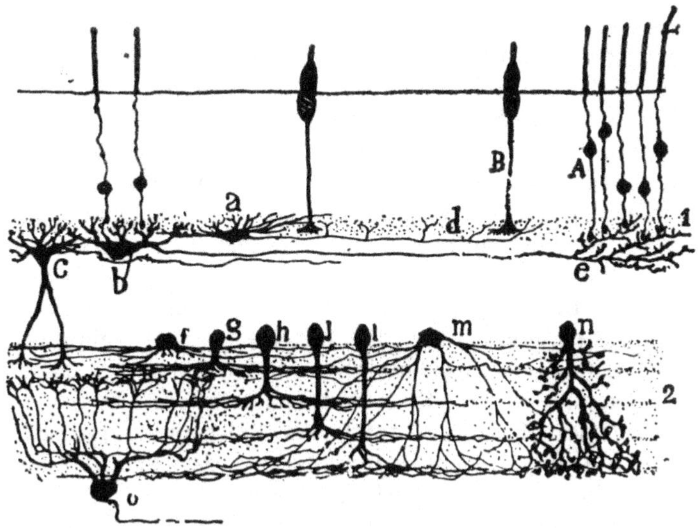

Fig. 1. Mammal retina: *A*, rods; *B*, cone; *a*, small horizontal cell; *b*, large horizontal cell; *e*, axon ending of horizontal cell (Cajal 1904)

made contacts through their dendrites with the rod's spherule and through the axon terminal with the "inferior endings" of undetermined photoreceptors (Fig. 1). Because of his ideas of centripetal axonal conduction, he felt confused when obliged to admit that the "visual impulse" originated in some receptors was transmitted through "these tangential neurons" and went back to photoreceptors located at some distance, forming a "vicious circle." In his own words: "In this way an important group of photoreceptors is lost for the conduction of the nerve impulse originated in the rods." To explain what Cajal thought was "nonsense," he supposed that "the HC do not represent authentic visual pathways but energetic centers related with the articulation established by the rods and the specific bipolars and with the function of reinforce the wave transmitted by the pair bipolar-ganglion cell." Cajal concluded that "probably the comparative histology will clarify the problem. In any case such system represents a fundamental retinianan factor because it has, even in invertebrates, an unusual richness and variety of forms."

Since Cajal's studies, advances in the knowledge of the retinal organization have occurred in the past 20 years. New cells were found, and the synaptic connections at the level of the two plexiform layers were clarified.

The first important contribution to the knowledge of the retinal structure since Cajal's studies was Polyak's (1941). He showed that the horizontal cells in the monkey retina were of a unique type, instead of the two varieties, the external and internal described by Cajal in the retina of low mammals. This fact, and his description of several types of bipolar, amacrine, and ganglion cells, led him to state that "the structural pattern of the retina is much more complex than conceived by Ramón y Cajal." In spite of this assertion, more than 20 years elapsed before new, important contributions on the structure of the retina were obtained.

The problem of the relation of horizontal cells with the "articulation" established by photoreceptors and bipolar cells was not solved by comparative histology as expected by Cajal. It took 60 years and electron microscopy to accomplish the task. Sjöstrand (1953) showed invaginations of the rod spherule, which were also described in the cone synaptic body by De Robertis and Franchi (1956), but it was Missotten (Missotten et al. 1953; Missotten 1965) who concluded that such invaginations enclose dendritic processes arising from the outer plexiform layer and coined the name "triad" for such compound structures. Structural details such as vesicles and synaptic ribbons, as well as contacts between cone synaptic bodies and cones and rod synaptic bodies were also described.

Three-dimensional reconstructions from electron micrographs of series of sections and the Golgi-EM technique have been the methods used to identify the processes included in the cone and rod synaptic complexes.

Serial sections were used initially by Sjöstrand in the guinea pig and by Missotten in the human retina. Missotten was able to identify the terminations of bipolar cells but not of the horizontal cells. Sjöstrand's approach of circuitry analysis through a three-dimensional study of a part of the outer plexiform layer was handicapped by an insufficient knowledge of the neurons, some of which were ignored in the analysis (interplexiform, axonless horizontal cell), despite their contribution with their processes to the outer plexiform layer neuropil.

More positive results were obtained by the Golgi-EM technique, that is, first identifying a Golgi-stained cell with light microscopy, then reincluding the isolated cell for ultrathin section and study with electronsmicroscopy. The electron-dense silver precipitate in the fine cell processes is readily identified in the sections, and the relationship of the processes to the synaptic bodies can be established. Using this technique, Stell (1965, 1967) was able to demonstrate that the lateral processes in the goldfish rod and cone synaptic complexes are the terminations of the horizontal cells. This technique has been of great help in studying the connections established by bipolar and horizontal cells and synaptic bodies of the photoreceptors.

Further contributions to the knowledge of the organization of the outer plexiform layer were obtained with the technique of silver impregnation of the entire retina studied in whole, flat-mounted preparations (Gallego 1953). This technique helped to find new cell types (associational ganglion cells, interplexiform cells) and, at the outer plexiform layer level, a new type of horizontal cell devoid of an axon, the axonless horizontal cell, initially named "amacrine cell of the outer plexiform layer" (Gallego 1964, 1965). Golgi staining of the entire retina was also a useful way to study the topographic distribution and structural features of Cajal's short axon horizontal cell.

The correlation between physiological and morphological observations has been clarified through the technique of intracellular recording of light-induced electrical activity followed by dye injection, since Kaneko (1970) demonstrated in goldfish retina that S potentials (Svaetichin 1953) were generated by horizontal cells.

3 Recent Structural Findings

The evidence obtained in the past 10 years has clarified the different types of visual and bipolar cells found in several animal species and has firmly established three new data of importance for the understanding of the functional organization of the OPL: (a) the transmission of the visual cell signals to bipolar cells is modulated by two systems of horizontal cells: the short axon and the axonless HC, different not only in their morphology but also in their connections with the synapses of photoreceptors with bipolar cells; (b) the existence at the OPL level of plexuses formed through "gap junctions" among the synaptic bodies of cones, the axonless HC, and, in reptiles and birds, among the axon terminals of the short axon HC; and (c) the existence of a feedback system between the two plexiform layers, mediated by the interplexiform cells.

3.1 Horizontal Cells

Two basic types of horizontal cells have been described in the tetrapoda retina (Gallego 1976a): the short axon horizontal cell and the axonless horizontal cell. The *short axon horizontal cell* is found in all tetrapoda retina. Described first by Cajal (1893) in low mammals, reptiles, birds, and amphibians, it was also studied in primates by Polyak (1941). Cajal (1893) classified this cell as a short axon neuron: "I believe that these cells are neural in nature and that they may be considered as neurons with a very short axon since they originate and finish within the retina itself."

Instead of the outer and inner, brush and stellate subtypes described by Cajal, modern studies have lead to the conclusion that only a single type of short axon horizontal cell exists in the tetrapoda retina but that it shows morphological differences according to the class and orders of tetrapoda and even in the same retina according to its location (central or peripheral). The differences refer to the number of dendrites and retinal area covered by them, to the axon length, and to the structure of the axon terminal. In all the retinas studied, the short axon horizontal cell dendrites form the lateral component of the cone triads, and its axon terminals penetrate mainly into the synaptic complex of rods, as has been shown in mammals (Kolb 1970, 1974; Fisher and Boycott 1974; Gallego and Sobrino 1975), birds (Gallego 1978; Tarrés 1982), and reptiles (Leeper 1978b).

The *axonless horizontal cell,* first described by Gallego (1964) in nonprimate mammal retinas, has also been found in the retina of birds (Gallego 1977; Mariani and Leure-Dupree 1977; Genis-Galvez et al. 1979), reptiles (Gallego and Pérez-Arroyo 1976; Leeper 1978a), and amphibians (Stephan and Weiler 1981).

The axonless HC build plexus which extend all over the retina, forming a well-defined layer. These cells establish wide membrane-to-membrane contacts, suggesting electrical coupling between them. The thin terminals which arise from the cell processes form the lateral components of the cone triads and do not have connections with the rod synaptic complex, as shown in mammals (Kolb 1974; Fisher and Boycott 1974) and birds (Gallego 1978; Tarrés 1982).

According to their connections, these two types of HC seem to have a different functional meaning: the short axon HC relates the cone-cone bipolar synapse with the rod-rod bipolar synapses, while the axonless HC is connected to only a single type of photoreceptor (cones in tetrapoda). Primates do not have axonless

HC and show instead a type II short axon HC (Gallego 1978) with the same connections with the photoreceptor synaptic bodies as the axonless cell has in low mammals (Kolb et al. 1980). The two basic types of horizontal cells, with few exceptions, are present in all vertebrate retinas.

Since their discovery in the last century, there has been some confusion about the HC nomenclature, particularly with the recent description of new cell types. It seems convenient to follow some rules in naming new cells, based not only upon their morphological features but also upon their functional connections. In our opinion, in all tetrapoda retinas we can consider, first, a division of HC into short axon and axonless types. In those retinas where more than one type of each basic HC is found, they can be named type I, II, etc, instead of using descriptive names such as stellate, brush, tufted, or candelabrumlike or names indicating their location, such as inner or outer, which in the past have been misleading. Considering their connections and their possible functional role, perhaps it would be more rational to name them on the basis of their relation to the synapse photoreceptor's bipolar cell. The short axon HC would be a cone-rod HC and the axonless HC a cone-cone HC. When the types of cones connected are known, they could be named accordingly (e.g., red-green cone HC).

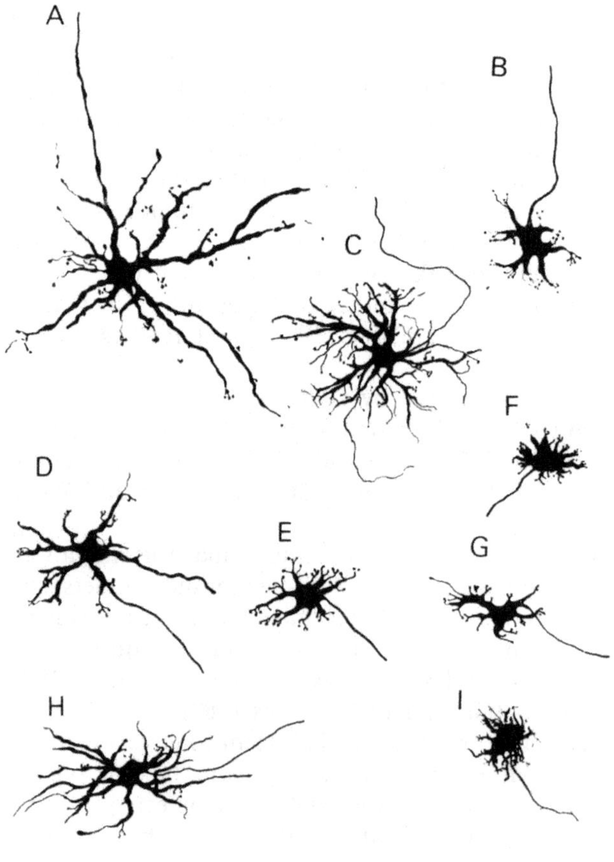

Fig. 2A – I. Short axon cell, somas: **A** primate peripheral HC; **B** primate parafoveal HC; **C** cat; **D** owl; **E** owlet; **F** chick; **G** eagle; **H** turtle; **I** frog

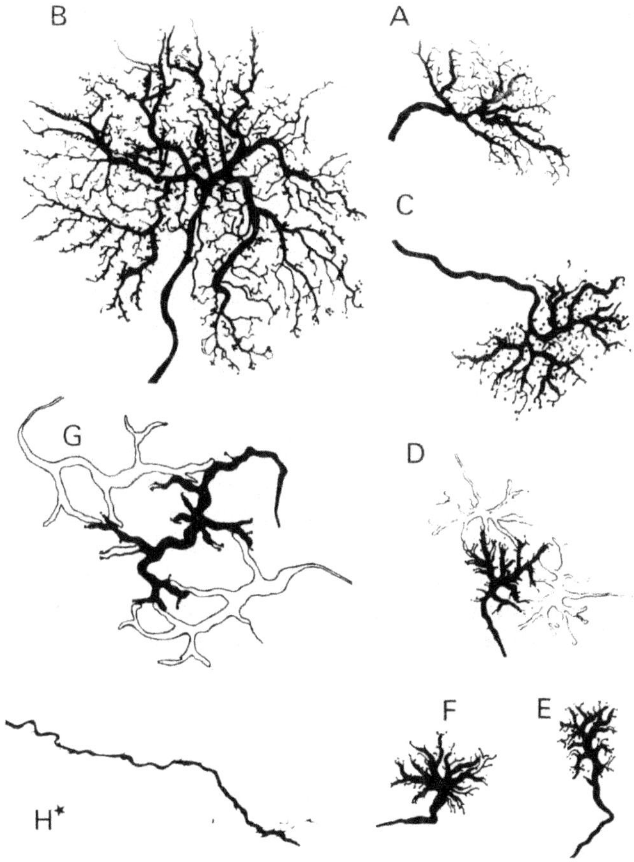

Fig. 3A – H. Short axon cell, axon terminals: **A** primate; **B** cat; **C** owl; **D** chick; **E** eagle; **F** owlet; **G** turtle; **H** frog (the frog ending is from Stephan and Weiler)

3.1.1 Short Axon Horizontal Cells

3.1.1.1 Primates. Two types of horizontal cells have been found in primates (Figs. 2a, b; 3a). The type I cell has been widely studied since its first description by Polyak (1941) in monkey retina. Near the fovea, it is a small cell with a body diameter of 8 – 10 μm, giving out five – eight dendrites directed towards the photoreceptor synaptic bodies. Its dendrites branch out in thinner processes which end up in round clusters of endings with small enlargements at their tips. Each cluster has a diameter of 6 – 8 μm, a size equal to or slightly smaller than that of the basis of the cone synaptic body. Golgi-EM studies have shown that each cluster endings form the cone triad lateral component (Kolb 1970). The area covered near the fovea by dendrites of this cell type is round with a diameter of 25 – 30 μm. The cell body gives out a 1 – 2 μm thick axon which runs straight, without branching, for a length of 1500 – 2000 μm. This axon widens into a thicker process which branches out several times. Thin processes arising from axon terminal branches run towards the photoreceptor synaptic bodies. These terminals end by knobs which penetrate the rod synaptic complex (Kolb 1970; Gallego and Sobrino 1975). The type I horizontal cell, located next to the fovea, connects with six to nine cones through its dendrites and with 350 – 500 rods,

some 1500 – 2000 μm away, through its axon terminal. Practically all the cones located in the parafoveal area covered by type I cell dendrites make contact with them. The areas covered by two neighboring type I cells do not overlap, although an occasional cluster terminal formed by dendrites from two cells can be seen.
The type I cell morphology changes progressively with the distance from the fovea. These changes apply mainly to the size and dendrite distribution, as well as to the area covered by them; the axon length and its terminals are similar to those cells close to the fovea. Dendrites of cells located 5 – 6 mm from the fovea, and especially the more distal ones, cover more irregular areas. These cells usually have one dendrite considerably longer than the others. The number of clusters formed by each cell, which determine the number of cones connected, increases with distance from the fovea. The peripheral cells show great overlapping of the areas covered by their dendrites, but each individual cell does not contact all the cones in that area.
The number of horizontal cells in the parafoveal zone has been calculated to be of the order of 1400/mm^2. This number decreases progressively with distance from the fovea. The axons of neighboring cells are randomly oriented in all directions, as can be seen in well-stained whole, flat-mounted retinas. A clear difference between the short axon horizontal cell and the axonless horizontal cell is that the former does not show membrane-to-membrane contacts among themselves, neither in their soma nor in their axon terminals.

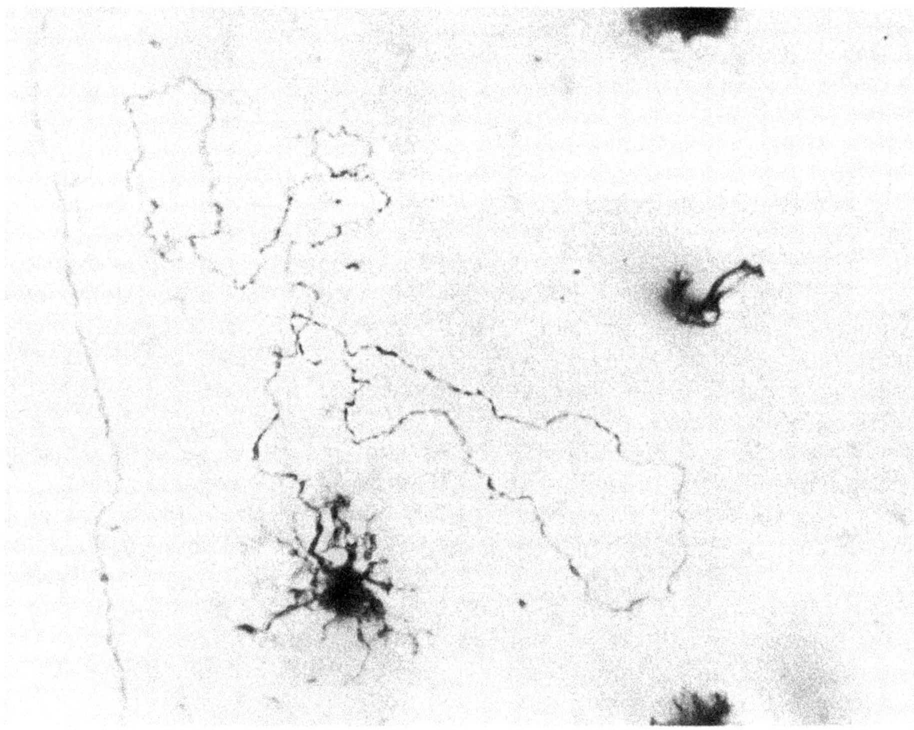

Fig. 4. Type II short axon HC of the monkey retina

The type II horizontal cell (Fig. 4) has recently been described in monkey (*Macaca mulatta* [Rhesus monkey]; *Macaca speciosa artoides*; *Cercopithecus*; *Erythrocebus*: Gallego 1977, 1978; Kolb et al. 1980) and squirrel *(Sciurus carolinensis)* retinas (West 1978). In monkeys, its structural features and functional connections are different from those of cell type I. Its cell body, 8 – 10 µm in diameter, gives out several very fine dendrites in a disorderly array, giving the cell a typical appearance. Its dendrite endings do not show terminal clusters but only one or two enlargements. A very thin short axon arises from the cell body, and instead of the straight course followed by the type II cell axon, it meanders around giving off some branches. The axon and its branches end by small ending clusters which, in Golgi-EM studies, have been seen to end up as lateral or central components of the cone triads. Out of the few cells studied, the dendrites of one of them, located at 1 mm from the fovea, made contact with ten cones, while those of another cell, at 4 mm from the fovea, made contact with 22 cones.
Type II cell connections are similar to those of nonprimate mammalian retina axonless horizontal cells. The area covered by type II cell dendrites varies with their distance from the fovea, but not as much as for the type I cell dendrites.

3.1.1.2 Low Mammals. The short axon horizontal cell was first described by Cajal in the retinas of cat, dog, rabbit, pig, sheep, and ox (Figs. 2C, 3B), having been also named in the past a "small" (Dowling et al. 1966) and an horizontal "B" cell (Fisher and Boycott 1974). It is similar to the type I horizontal cell of the primate retina. The type II short axon horizontal cell, recently described in the monkey retina, has been only found in the squirrel, an all-cone retina (West 1978), among low mammals.
In the cat retina, where type I HC have been widely studied, they do not show the morphological differences described for the primate: due to the dispersion of their dendrites, only slight differences can be seen between the central area horizontal cells and the more peripheral ones. The cat retina does not have short axon HC with the morphology of the primate retina parafoveal HC, but rather resembles the intermediate or peripheral primate cell.
In the cat and dog the horizontal cell body diameter is about 10 µm and gives out 8 – 12 dendrites which branch out dichotomously two or three times. These branches show thin terminal fibers which end in small clusters of knobs. The number of knobs in each cluster is lower than in the primate retina, because the number of triads in the cone synaptic bodies is smaller in cat than in primate retinas. The cell dendrites cover a circular area 100 – 140 µm in diameter, and 35 – 45 clusters can be seen in each cell, which means that it connects with the same number of cone pedicles. Dendrite endings form the cone triad lateral component, as has been observed in Golgi-EM studies (Kolb 1974; Fisher and Boycott 1974).
When one horizontal cell is completely stained, the main differences with the same type of primate retina can be seen in whole, flat-mounted preparations. These are: (a) the lesser length of the axon, being usually no longer than 400 µm; (b) the larger size of the axon terminal arborization which covers retinal areas of the order of 150 × 250 µm; and (c) the number of knobs in a terminal area of the same size is larger than in the primate retina.
Golgi-EM studies of the axon terminal have shown (Kolb 1974; Fisher and Boycott 1974) that the endings penetrate into the rod synaptic complex. The number

of rods connected through one axon terminal has been estimated to be on the order of 3000 – 4000. This number of rods is connected by one HC to 35 – 45 cones located 400 μm away.

3.1.1.3 Birds. Cajal (1893) achieved a complete staining of a short HC in the sparrow retina which he named brush HC. In all diurnal birds, the axon ends in a wide enlargement (Figs. 2D – G, 3C – F). When the axon terminal is stained independently, it looks like an individual cell, a fact that mislead Cajal into considering this structure a different cell type, his stellate HC (Gallego et al. 1975b). Studies on the chick embryo retina confirm this interpretation of Cajal's stellate cell (Gallego 1982).

An interesting observation is the structural difference of the OPL between diurnal and nocturnal birds. Cajal showed that the visual cell synaptic bodies of diurnal birds are located at three different levels in the OPL: double cones and rods at the most scleral level, straight cones at the intermediate level, and oblique cones at the inner or vitreal level. In a bird such as the owlet which hunts at dawn, or in a purely nocturnal bird such as the owl, all the visual cell synaptic bodies lie at a single level in the OPL (Gallego et al. 1975a; Tarrés 1982). At the same time, striking differences in the morphology of the short axon HC have been shown.

In diurnal birds and in whole, flat mounts of Golgi-impregnated retinas, this cell type shows a small body (10 μm in diameter), and axon 200 – 800 μm long, depending on the avian species, and the axon terminal. The cell body gives out five to eight dendrites that run towards the synaptic bodies of the photoreceptors. These dendrites branch out a couple of times and end by small knob clusters, similar to those found in the HC of the mammalian retina. Each cluster is formed by four to six knobs, covering an area slightly smaller than the one covered by a cone's synaptic body (6 μm diameter). The total area covered by dendrites from one horizontal cell has a diameter of 30 μm. The axon, 1 – 2 μm thick, is 100 μm long in the newborn chick, 300 – 400 μm in the hen, 600 – 800 μm in the falconidae, and 700 – 900 μm in the owlet retina. In all the retinas of diurnal birds, the axon ends in a wide enlargement which gives out thin terminal fibers (Gallego et al. 1975b). The axon terminal ramifies into thinner branches which spread out giving rise to delicate endings that run towards the photoreceptor synaptic bodies and end in isolated knobs without forming clusters. These knobs are larger than the ones found in the dendrite clusters.

Cross-sections of the retina show the short axon cell bodies located in the outer row of nuclei of the inner granular layer. The dendrites, directed toward the photoreceptor synaptic bodies, end mainly at the outer row of synaptic bodies in the plexiform layer, but very often, dendrites directed towards the straight and oblique cones can be seen (Gallego 1976a).

The axon endings are located above the cell body, as well as below the first row of synaptic bodies intermingled with the synaptic bodies of the straight and oblique cones. Their thin terminal fibers end by thick knobs at the level of the external row of synaptic bodies.

The cell body and terminals of the short axon HC have been identified by electron microcopy (Gallego et al. 1975; Barón and Gallego 1982). The observations on the axon terminals are of great interest; they form a plexus with membrane-to-membrane contacts among them. These contacts are of the gap junction type, but they consistently show vesicles at either side of the membranes.

Golgi-EM studies have shown that the short axon HC dendrites establish contact with the double cones of the outer row, the straight cones of the middle row, and with the oblique cones. The axon terminals penetrate the rod and some single cone synaptic complexes (Gallego 1978; Tarrés 1982).

In a seminocturnal bird such as the owlet, both the soma and axon terminal of the short axon HC have the morphological features of the diurnal birds, only the axon is longer (800 μm).

The short axon HC of the owl, a nocturnal bird with a greater proportion of rods in his retina, shows some morphological differences if compared with the same cell type of diurnal birds. The soma gives out three of five long and thick dendrites which are reminiscent of the same cell type located at 6 – 8 μm from the fovea in the monkey retina. The axon length has not been determined, but with the observations we have up to now, a length superior to 800 μm can be suspected. Where the differences with the diurnal birds are striking is in the structure of the axon terminal: it is identical to the axon terminal in the retina of primates and probably does not form a plexus, as the same structures do in diurnal birds and reptiles.

3.1.1.4 Reptiles. Two types of short axon HC were described by Cajal (1893): brush and stellate. Cajal's description of the brush HC in the lizard (Figs. 2H, 3G) was limited to the body, dendrites, and the beginning of the axon; he was not able to follow the axon to its terminal nor to stain it. His description of the stellate cell in all details, except for the presence of an axon, corresponds to the axonless HC which was clearly described many years later.

The axon terminal of the turtle brush HC was first seen in silver impregnations of whole, flat-mounted retinas (Gallego and Pérez Arroyo 1976). These terminals form a plexus located between the photoreceptor synaptic bodies and the horizontal cells' somas. EM studies clearly show this disposition, as well as the existence of abundant specialized membrane contacts between the axon terminals. The axon, 300 – 400 μm in length, ends without branching in a tuberous structure which gives out short, thick processes leading to thin terminal fibers directed toward the photoreceptor synaptic bodies. Each terminal fiber shows at its tip one or two small knobs.

The soma is located at the outer row of cells of the inner granular layer and has a diameter of 10 – 12 μm. The outer pole cell gives out several thick dendrites which show thin terminal fibers directed toward the visual cell synaptic bodies and which end by a small group of knobs (Gallego and Pérez Arroyo 1976; Leeper 1978a).

This cell type morphology varies with its location in the retina: it is smaller near the visual streak, and its dendrites cover eliptical areas on the order of 30 × 150 μm (Norman and Kolb 1981). The peripheral cell dendrites spread more horizontally and cover larger areas of the retina.

Golgi-EM studies have shown that the dendrites contact with double cone synaptic bodies (Pérez Arroyo 1978), red cones, and green cones, and the axon terminals with red cones and rods (Leeper 1978b).

3.1.1.5 Amphibians. Two types of short axon HC were found by Cajal (1893) in the frog retina (Fig. 3H, I): outer and inner HC. In agreement with Stephan and Weiler (1981), our own studies show that, as is true in all tetrapoda retina, only one type of Cajal short axon HC can be found in the frog.
In Golgi-stained whole, flat-mounted retinas, the soma size is not easy to measure due to the image given by the short and thick dendrites, packed together, directed towards the photoreceptor synaptic bodies. From the dendrites tip arise thin fibers which end in very small knobs. The retinal area covered by the dendritic tree has a diameter between 40 and 90 μm. The axon, very thin, has a length between 200 and 800 μm, follows a usually straight course, and ends by a slight swelling which has a diameter of 3 μm and a length, difficult to determine, of about 30 μm (Stephan and Weiler 1981). As detected first by Cajal (1893), along its course it gives out short and thin collaterals directed towards the photoreceptor synaptic bodies.

3.1.2 Axonless Horizontal Cells

3.1.2.1 Low Mammals. The axonless HC was first described by Gallego (1964, 1965) as an amacrine cell of the outer plexiform layer in the retina of low mammals (Fig. 5A). It was subsequently studied by electron microscopy (Fisher and Boycott 1974; Sobrino and Gallego 1970; Gallego 1971), in Golgi-stained retinal sections (Gallego 1975, 1971), and in Golgi-stained, whole, flat-mounted retinas

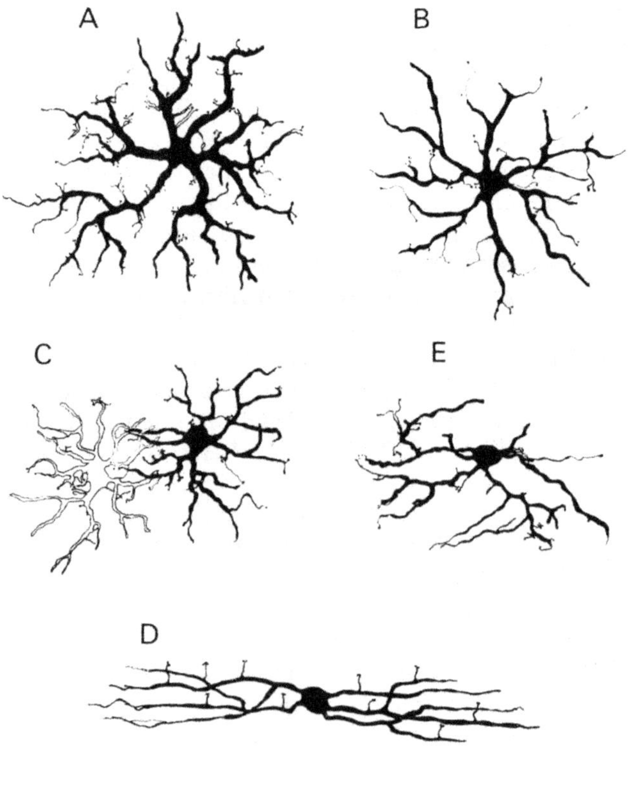

Fig. 5A – E. Axonless HC: **A** cat; **B** turtle; **C** eagle type II; **D** eagle type I; **E** frog

(Dowling et al. 1966; Gallego 1975). It was also called large (Dowling et al. 1966) or horizontal A cell (Fisher and Boycott 1974).

The technique of whole-retina silver impregnation (Gallego 1953) shows these cells forming a single layer which extends all over the retina, including the area centralis, in cat and rabbit. The neurofibrillar staining methods, especially Balbuena and Cajal's reduced silver methods, stain this cell type very easily. On the contrary, the rapid Golgi technique used by Cajal infrequently stains it, which may explain why it was not noticed by Cajal in his study of the vertebrate retina.

The number of cells varies with the retinal area (Orellana and Gallego 1959): in the cat it was found to be 100 – 140 cells/mm^2 in peripheral retina, 250 – 300 cells/mm^2 in the juxtapapillar zone, and 500/mm^2 in the area centralis.

By using whole, flat-mounted retinas stained by the Golgi method, these cells can be seen isolated. Their cell bodies have diameters of about 10 – 14 μm, and give out several cell processes (Gallego 1975; Kolb 1970), spreading horizontally and branching dichotomously. At their origin these processes are as thick as the cell body, making it difficult to delineate the cell body morphological limits. These processes give out thin fibers directed towards the photoreceptor synaptic bodies. These fibers end in a very small cluster of knobs or in a single knob. The number of knobs in each cluster is not greater than four. The retinal area covered by the cell processes varies with their location. At the periphery, the area covered by a single cell is practically circular, 150 – 200 μm in diameter; in the area centralis the diameter is only 50 μm. Great overlapping of neighboring cell areas is always observed.

This cell type contacts the cone synaptic bodies, its terminal knobs forming the lateral component of the cone triad (Kolb 1974). Wide membrane-to-membrane contacts between neighboring cells can be observed, suggesting electrical coupling among them.

3.1.2.2 Birds. Two types of axonless horizontal cells (Fig. 5E, D), called types I and II, have been described in the avian retina (Gallego 1978, 1977; Genis-Galvez et al. 1979).

The axonless horizontal cell type I is a very characteristic cell of the avian retina (Gallego 1977, 1978). It has a round body, 7 – 9 μm in diameter, which gives out five to seven processes in an horizontal plane from opposite poles, giving the cell a bitufted aspect. The cell processes branch a few times and are 50 – 60 μm long, expanding through a width of 25 – 30 μm; that means that the area covered by the dendrites is a 120 × 25 μm rectangle. These processes give out very thin terminal fibers which, isolated and without branching, reach the outer row of the photoreceptor synaptic bodies. Each thin terminal ends in a small cluster formed by no more than two to four knobs. Each cell shows from 16 to 24 clusters, and each one connects with one cone which means that each cell has contact with 16 – 24 cones in a 120 × 25 μm rectangular area. The distance between connected cones is from a minimum of 5 μm to a maximum of 20 μm.

The terminal knobs are located at the outer row of photoreceptor synaptic bodies, as can be seen in cross-sections and in whole, flat-mounted retinas. The endings are located at the same level as the short axon horizontal cell dendrites, clusters, and axon terminal.

The processes of neighboring cells intermingle, and because they are bitufted, they look as if they form parallel rows.

Golgi-EM studies have shown that the terminal knobs of these cells penetrate in the double cone synaptic complex (Gallego 1978).

The axonless horizontal cell type II has recently been fully described (Gallego 1978, 1977; Mariani and Leure-Dupree 1977; Genis-Galvez et al. 1979), although it was already noticed by Cajal, who drew it (Cajal 1888) in a cross-section of the retina and made a brief description of it under the name of stellate or subepithelial. Nevertheless, Cajal (1893) subsequently described the brush horizontal cell and its axon terminal in the avian retina, but thinking that this terminal was an independent cell, he also named it stellate horizontal cell. (Looking at Cajal's drawings, it is clear that the stellate cell of his 1888 paper is completely different from his stellate cell of the 1893 study.)

The cell body of 8 – 10 μm diameter gives out three to four thick processes which branch out several times, covering a circular area of 100-μm diameter. These cell processes give out thin and short terminal fibers, directed towards the photoreceptor synaptic bodies. The thin terminal fibers end in a small cluster of two to four knobs. These knobs are located below the short axon cell and axonless type I endings, at the intermediate row of the outer plexiform layer.

Preliminary Golgi-EM studies have shown that the terminal knobs penetrate in the synaptic complex of the straight cones containing the red oil droplet (Gallego 1978).

3.1.2.3 Reptiles. As in the avian retina, two types of axonless horizontal cells have been found in reptile retina (Fig. 5B), type I and type II. The type I cell, similar to the axonless type I in the avian retina, has not been easily stained. In Golgi stainings (Pérez Arroyo 1978), it appears as a bitufted cell with a round body (7 – 9 μm in diameter) which gives out several processes (from three to five) in an horizontal plane at each opposite pole. Thin terminal fibers come out from the processes running towards the photoreceptors. These fibers end in a small cluster of knobs. In the turtle retina, the distinction between the short axon cell and the axonless type I is not as clear as in the avian retina. Nevertheless, the stained cells obtained allow the axonless cell type I to be considered as an independent cell type, similar to the axonless type I cell of the avian retina. Further studies in the reptile retina are needed to determine cell connections with the photoreceptors.

The axonless horizontal cell type II is similar to the axonless type II horizontal cell of the avian retina and has been clearly seen in Golgi-stained turtle retina preparations (Gallego and Pérez Arroyo 1976), probably being the stellate cell of Cajal (1893, plate III, Fig. 7) who erroneously stated that one of its processes was an axon.

It has a round body, 9 – 12 μm in diameter, which gives out four to five large horizontally extending processes at its scleral pole. They split dichotomously into thinner branches which cover approximately circular areas of about 100 μm in diameter. A small number of thin fibers arise from these processes, run toward the photoreceptors, and end in a small cluster of three to five knobs at the photoreceptor synaptic body level.

3.1.2.4 Amphibians. The axonless HC has been recently described as HC in the frog retina (Stephan and Weiler 1981). We have also found this cell type recently

(Fig. 5E). It has a 10 μm diameter soma which gives out three to six processes, horizontally spread. The cell processes split into thinner branches and cover a near circular area of 70 – 160 μm in diameter. From the processes emerge thin fibers which end in a cluster of very small knobs at the level of the photoreceptor synaptic bodies. The connections of the axonless HC are not known yet.

3.2 Cell Plexus in the Outer Plexiform Layer

Plexus formed by membrane-specialized contacts between visual cells, axonless HC, and axon terminals of the short axon HC have been found in several tetrapoda retina.

3.2.1 Plexus Formed by Photoreceptor Synaptic Bodies

Contact between synaptic bodies of photoreceptors were first revealed in the guinea pig (Sjöstrand 1953, 1958) and found afterwards in all tetrapoda retina studied. Two kinds of contacts have been described in mammal retina: between cones and between rods and cones (Fig. 6). Contacts between synaptic bodies of rods are not known in mammal retina but have been found in mud puppy and toad retinas (Custer 1973; Fain et al. 1976).

Several rod synaptic bodies are in contact through tiny gap junctions with the synaptic body of a single cone, mainly through its basal filaments in mammal retina. This observation gives relevance to Balbuena's old and forgotten concept (1936) of the organization of visual cells into "photoreceptor units" formed by one cone and several rods, five to ten in primate retina, according to our own studies. This type of contact was initially considered as a cleavage point, without functional significance, but there is now direct evidence that in cat retina the cones have mixed rod and cone input (Nelson 1977) with the deduction that the gap junctions between rods and cones may provide the pathway by which the rod signals enter the cones. The possibility that cone signals enter rods through the same mechanism must be considered.

The synaptic bodies of the cones are connected between them through the same type of tiny gap junctions, mainly through their basal filaments found in several species including the turtle and the macaque (Raviola and Gilula 1973). In this way, transverse propagation of the cone activity through their synaptic bodies can occur, as demonstrated by signal sharing between cones found in the turtle (Baylor et al. 1971).

The signal sharing between rods found in snapping turtle and toad retinas (Copenhagen and Owen 1976) is explained by the gap junctions between synaptic bodies of rods found in these species.

3.2.2 Plexus Formed by the Axonless Horizontal Cells

The axonless HC of cat and other low mammals are arranged in a single and well defined layer which extends all over the retina, including the area centralis (Gal-

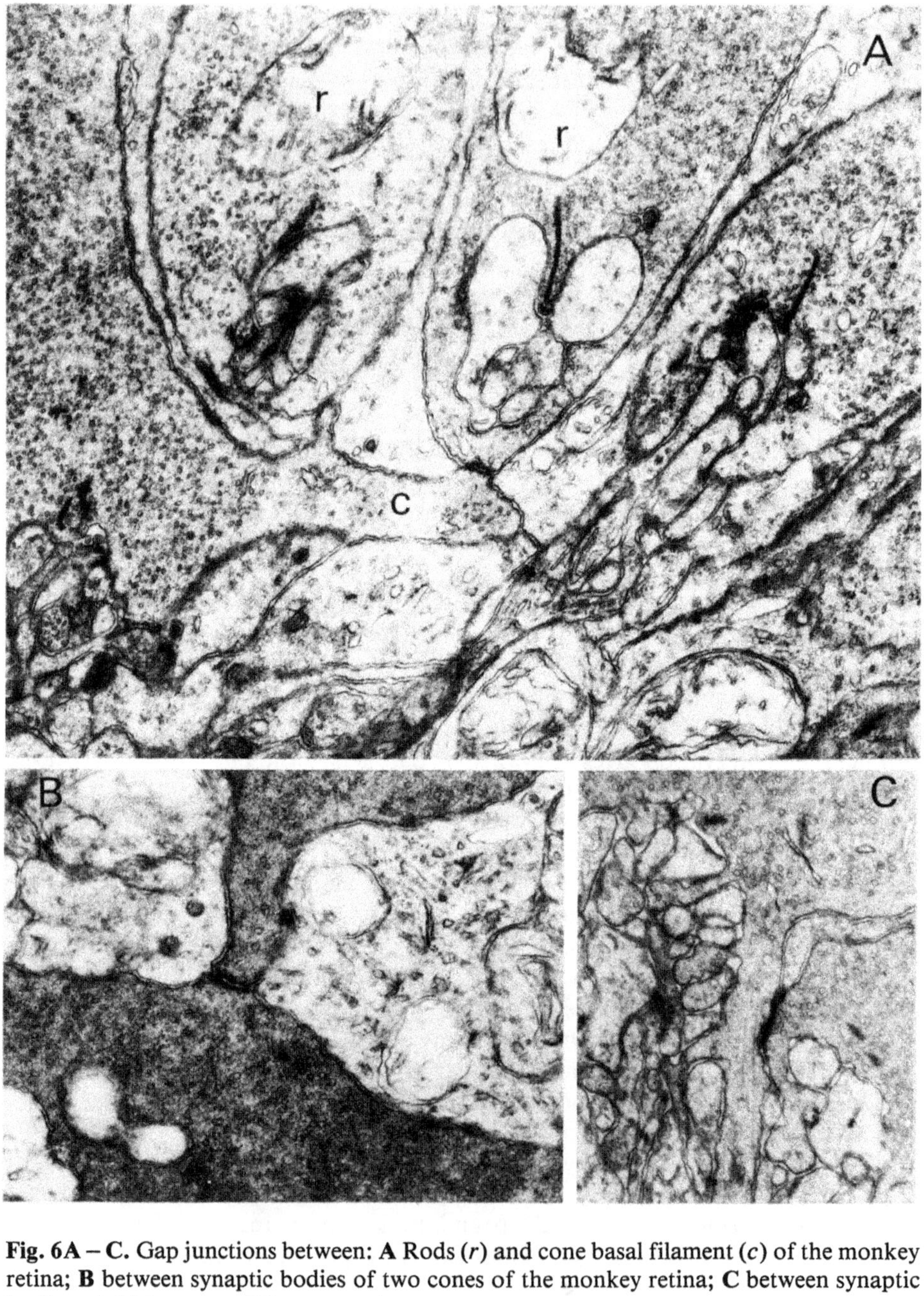

Fig. 6A – C. Gap junctions between: **A** Rods (*r*) and cone basal filament (*c*) of the monkey retina; **B** between synaptic bodies of two cones of the monkey retina; **C** between synaptic bodies of chick cones of the external row

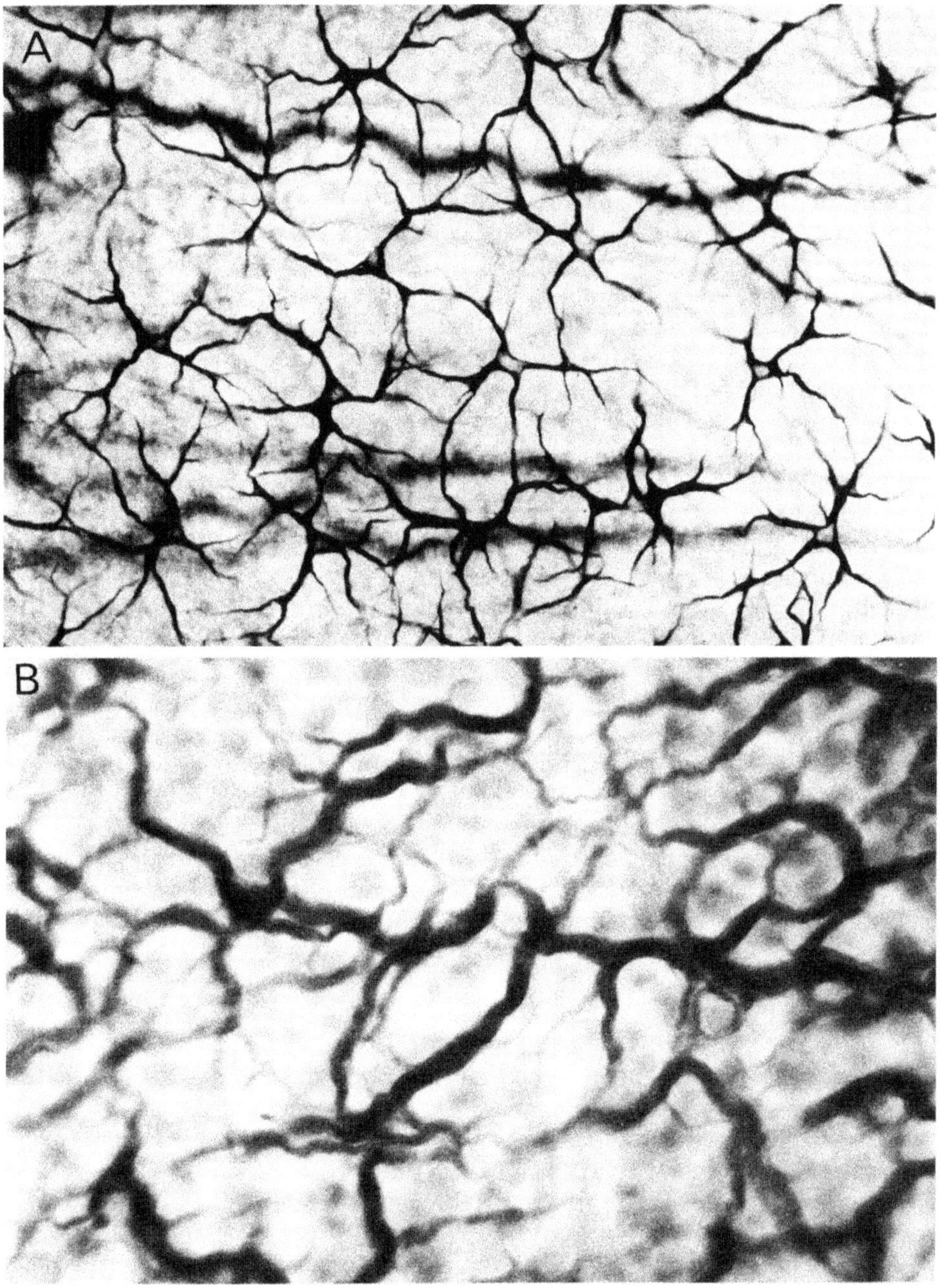

Fig. 7. A Cat axonless HC plexus; **B** turtle axon terminal plexus

lego 1964, 1965). Electron-microscopy studies revealed that the axonless HC form a syncytium through gap junctions between their processes as shown in cat (Sobrino and Gallego 1970; Kolb 1977) and rabbit (Raviola and Gilula 1976). This plexus is similar to the ones formed by HC in fish retina and known since the last century (Fig. 7A).

In teleost retina, light-induced S. potentials spread laterally and with decrement to neighboring nonilluminated areas (Mitarai et al. 1961). Specialized membrane contacts between HC were supposed to be the agent for these enlarging HC receptive fields (Svaetichin et al. 1965). In the same way the receptive field of cat axonless HC exceeds the retinal area covered by its cell processes (Nelson 1977), which implies a similar mechanism of lateral spreading of activity through gap junctions interpreted as functional connections between them. The different types of axonless HC found in submammalian species probably build similar syncytia, but solid evidence in favor of this possibility is lacking.

In mammals, evidence of specialized membrane contacts between somas or dendrites of the short axon HC has not been found. Among tetrapoda, only reptiles and diurnal birds show plexus formed by the axon terminals, which in mammals and nocturnal birds appear without any kind of specialized membrane contacts between them.

The morphological observations emphasize an important difference between the two types of HC in mammals: the short axon cells are individual neurons relating the cone-cone bipolar synapses with the rod-rod bipolar synapses, and spreading of activity between these cells is not to be expected, while the axonless HC connected only to cone-cone bipolar synapses, and forming a syncytium, gives the structural basis for the spreading of activity between them as shown by electrophysiological studies.

In contrast with the evidence obtained in mammals, the plexus formed by the axon terminals in reptile and diurnal bird retinas is a good example of the differences of organization found among species and of the risk of extrapolating to others the data obtained in one particular retina.

3.2.3 Plexus Formed by the Axon Terminals of the Short-Axon Horizontal Cells

Among tetrapoda, only reptiles and diurnal birds show plexus formed by the axon terminals (Fig. 7B). In mammals they have not been detected, and we do not have studies in amphibian retina.

In turtles (Gallego and Pérez Arroyo 1976) and in diurnal birds (Gallego 1976), the axon terminals of the short axon HC form a well-defined plexus located between the visual cell synaptic body and the scleral row of the inner nuclear layer formed by the HC somas. This diposition make the axon terminals the most frequently impaled structures in electrophysiological studies of the turtle retina. Electron-microscopy studies show very wide gap junctions between these terminals in turtle (Pérez Arroyo 1978) and birds, where a frequent finding is vesicles on both sides of the junction (Barón and Gallego 1982).

Two subtypes of the HC were initially detected through intracellular recording of light-induced responses: a large receptive field type (L1) and a small receptive field type (L2) (Simon 1973). Comparing the images obtained by procion yellow

marking of these cells and Golgi stainings, it was concluced that the two luminosity responses, L1 and L2, arise from the same cell: L1 from the axon terminals and L2 from the bodies of the short axon HC (Leeper 1978a).
There is physiological evidence showing that the large field of an L1 cell is due to activity of the neighboring axon terminals interconnected through gap junctions (Piccolino et al. 1982).

3.3 Interplexiform Cells

The combination of histological and immunofluorescence techniques has allowed the detection of new cell types which were previously included in the general classification of amacrine or bipolar cells. Some of them are actually subtypes of amacrine cells, such as the indoleamine-accumulating cells (IA) described in the retina of some vertebrates (Ehinger and Floren 1976), which mutually contact other amacrine and bipolar cells in the inner nuclear layer or the dopamine-containing neurons originally named "adrenergic junctional cells" (Ehinger 1966). Neither cell type has been found to send processes to the OPL.
A different, new cell type, with its soma located at the vitreal row of the inner nuclear layer and processes spreading to both plexiform layers, was first described in Golgi preparations of the cat retina and named "interplexiform cell" (Gallego 1971). It is also found in many vertebrate retina (Boycott et al. 1975). Histofluorescence studies had previously revealed the presence of dopamine-containing neurons (DA) in several vertebrate retinas (Ehinger and Falk 1969). Some of these DA neurons, which have processes extending to both plexiform layers, were only found in the retina of *Cebus* monkey and various species of fish (Dowling and Ehinger 1975; Neghisi and Drujan 1978) and were actually interplexiform cells. In most mammals, however, interplexiform cells are not dopaminergic.
Another subtype of interplexiform cell has been recently detected (Marc and Lam 1981). It is a glycine-accumulating neuron with processes distributed in both plexiform layers and could mediate the glycine-evoked responses detected in the HC.
The DA interplexiform cells in *Cebus* are presynaptic to amacrine somas and processes. They receive synapses from amacrine processes in the inner plexiform layer and are presynaptic to both bipolar and HC at the OPL level (Dowling et al. 1980). According to these contacts, the interplexiform cell provides a feedback pathway from the inner to the outer plexiform layer, as was suspected (Gallego 1971).

4 Correlation Between Morphological and Functional Observations

Intracellular recording followed by the cell marking with dyes and cell identification according to data obtained by histological studies have been the procedures employed to relate functional and morphological characteristics.
The photoreceptor response is a hyperpolarizing potential (Werblin and Dowling 1969), modified along the retinal pathway through cell interactions at the level of

both plexiform layers. The response latency and rise time is twice as fast in cones as in rods, as shown in several submammalian species – mud puppy, turtle, and frog (Fain 1975; Baylor and Hodgkin 1973; Toyoda et al. 1970), and cat (Nelson 1977) – a property used to detect both inputs in HC recordings.

Turtle cone response to light depends upon direct stimulation of the recorded cell and also upon stimulation of neighboring cones (Baylor et al. 1971). Direct interaction at short distances results in a summation of responses and is electrically mediated through tiny gap junctions between cones. Cat cones have large receptive fields, with space constants of 50 – 190 μm, due to their synaptic bodies coupling through gap junctions and are strongly influenced by rods. Rod influence in the cone response (Nelson 1977) may be due to the constant finding of photoreceptor units formed by one cone synaptic body connected to several rod synaptic bodies through tiny gap junctions. This structural organization may explain the failure to detect a response purely driven by these photoreceptors in the axonless HC, which is connected only to cones.

Recording of pure rod input in the axon terminals of the short axon HC (Steinberg 1971; Niemeyer 1975; Nelson et al. 1976; Foerster et al. 1977) supports the concept that cone influence on the rod response is negligible. The difference in the signal sharing between cones and rods can be explained (Molenaar 1981) because several rods send their signal through gap junctions to one single cone, whereas each rod is contacted by only one or two cones.

Intracellular recording in the dark has revealed that cones which show no signal sharing with other cones have a high level of random noise in the membrane potential (Simon et al. 1975) and that photoreceptors in turtle that share signals over large retinal areas exhibit low levels of membrane noise in the dark (Lamb and Simon 1976). As a consequence, a possible function of photoreceptor coupling might be the reduction of noise to improve the signal-to-noise ratio.

The first intracellulary recorded responses in retina were the S potentials (Svaetichin 1953) generated by HC. They were again demonstrated several years later in fish and cat retina (Kaneko 1970; Steinberg and Schmidt 1970). As shown in fish retina, the S potentials are two types of graded responses to illumination: L potentials which hyperpolarize to all light wavelengths and C potentials which show hyperpolarization to short wavelengths and depolarization to long wavelengths (Svaetichin and McNichol 1958). It was suggested that the function of L potentials was to code luminosity, and the function of C potentials was to code color. The response of HC to illumination of a restricted area spreads to neighboring nonilluminated areas (Mitarai et al. 1961). It is widely accepted that the lateral spread depends on electrical coupling between HC (Naka and Rushton 1967; Kaneko 1971).

Functional and morphological observations show differences in both types of HC in several animal species. The morphology and synaptic connections of the short axon HC varies among tetrapoda according to the photoreceptor types found in their retinas. The differences are striking in the morphology of the axon terminal between mammals and submammalian species (reptiles and birds). In mammal retinas which contains rods and cones, the area covered and number of rods contacted by the axon terminal is different between low mammals and primates in relation with the proportion of rods in each retina. In the all-cone squirrel retina, the axon terminal, a structure related only to rods in mammals, has not developed; a type II short axon HC whose axon endings contact cones has been found insteadt (West 1978). Among birds, the owl retina, which is very rich in rods, shows axon terminals similar to the same structure in the primate retina, being morphologically very different from the diurnal birds.

In reptile and diurnal bird retinas, the axon terminals form a syncytium, while in mammals (and nocturnal birds such as the owl) they are not connected to each other – an observation to be considered when analyzing the electrophysiological observations made in different species.

The morphology and synaptic connections of the axonless HC are similar in all tetrapoda retina studied. Its cellular processes form the lateral component of the cone triads and do not have direct connection with the rod synaptic bodies. The axonless HC form well-defined plexus in low mammal retinas, similar to the ones formed by the axon terminals in bird and reptile retinas. The axonless HC has not been found in primates; their retinas have, instead, a type II short axon HC connected only to cones, which could be functionally analogous to the sub-primate axonless HC.

Due to their different connections with photoreceptors, it was expected that the two basic types of HC would exhibit different types of S potentials. This supposition proved to be correct in some retinas but was not generalized to all species studied. For example, in the turtle and the cat where the two types of S potentials have been recorded, the short axon HC gives L potentials in both retinas, but the axonless HC gives C potentials in turtle and L potentials in the cat.

There is a great amount of experimental evidence on S potentials in turtle retina where both types of responses, luminosity (L) and chromaticity (C), have been recorded. The short axon HC is the source of two L responses (Leeper 1978a) which can be distinguished by their receptive fields (Simon 1973): a large-field HC (L1 HC) originates at the axon terminal and a small-field HC (L2 HC) originates at the soma. Except for the receptive field sizes, the L1 HC and L2 HC responses were the same to bright stimuli; however, they do not contact the same types of photoreceptors. It was found that L2 HC contacted red- and green-sensitive cones and double cones, whereas L1 HC did not contact green-sensitive cones but double cones, red-sensitive cones, and rods. In dim light, L1 HC showed a rod-dominated response (Leeper and Copenhagen 1979), but brighter stimuli gave a fast hyperpolarization due to red cone input in both L1 HC and L2 HC.

The differences between the receptive field size of L1 HC and L2 HC have been explained by structural evidence. The existence of wide gap junctions between the axon terminals (L1 HC) and the observation that Lucifer yellow injected into one L1 HC diffuses to an extensive network of axon terminals explain the larger receptive field of this structure, if compared with the receptive field of L2 HC which do not show gap junctions between them. GABA, a neurotransmitter found in the retina, is probably involved in the function of gap junctions between axon terminals, as suggested by the observation that bicuculline and picrotoxin, two GABA antagonists, narrow L1 HC receptive field and prevent diffusion of Lucifer yellow injected into one terminal (Piccolino et al. 1982).

The soma (L2 HC) and the axon terminal (L1 HC) of the short axon HC are connected by an axon about 300 μm long, but there is experimental evidence in turtle retina (Simon 1973) showing that both structures function independently, a surprising property in a neuron.

HC with C responses, hyperpolarized to green light and depolarized to red light (R/G HC), were initially found (Miller et al. 1974) followed by the finding of another HC type which hyperpolarizes to blue light and depolarizes to green light (G/B HC) (Fuortes and Simon 1974). The sources of these C responses were two axonless HC which are nearly identical morphologically (Leeper 1978a). A possibility is that the two C responses (R/G and G/B) arise from two morphologically different subtypes of axonless HC (Pérez Arroyo 1978).

Several authors have produced physiological evidence of feedback in chemically mediated synapses between HC processes and cone synaptic bodies (Baylor et al. 1971; Fuortes 1972; Simon 1974), although the morphological characteristics of chemical synapses between the two structures have not been clearly defined. This feedback mechanism explains the antagonistic interaction found between cones located at distances of more than 300 µm, and it is probably mediated by HC plexus, either of the axon terminals or of the axonless HC, or both.

In the cat retina, the short axon HC and the axonless HC give both the L type of S potentials (Grusser 1960) and receive mixed rod cone input (Brown and Murakami 1978; Steinberg 1969). All HC recorded showed receptive fields larger than the area covered by their dendritic trees or cellular processes (Steinberg 1969; Foerster et al. 1977; Nelson 1977; Molenaar 1981).

Records have been obtained from three sources structurally identified as short axon HC body, short axon HC terminal, and axonless HC (Nelson et al. 1975). The pure rod input found in some of these structures (Steinberg 1971; Niemeyer 1975; Nelson et al. 1976; Foerster et al. 1977) probably corresponds to the short axon HC terminal, only connected to rods. A mixed rod-cone HC response was detected in both the soma of the short axon HC and the axonless HC, structures connected only to cones (Nelson et al. 1975). The rod contribution to the L potential recorded from axonless HC can only be explained by the rods contribution to the cone signal through gap junctions between their synaptic bodies. The same explanation can be admitted for the mixed rod-cone responses obtained from the soma of the short axon HC, although the rod signal detected could propagate from the axon terminal, an unlikely possibility (Nelson 1977).

In the cat short axon HC, as in the same cell type of the turtle, it has been postulated that the physical properties of the axon do not allow the spread of electrical signals between soma and axon terminal (Nelson et al. 1975), but the possibility that the axon conducts signals from the soma to the axon ending must not be excluded, as it is the only way to explain that stimulation of cones at photopic intensities reduces or abolishes the rod receptor potential (Whitten and Brown 1973).

The short axon HC soma has a smaller receptive field than the axonless cells. In both structures the size of the receptive field is larger than the area covered by their cell processes. The larger receptive field of the axonless HC is explained by spread of activity through the plexus built by these cells. It is difficult to accept that the same mechanism explains the receptive field size of the short axon HC somas, larger than the area covered by its dendrites, because specialized membrane contacts between them have not been found. The influence on the field size by lateral spread of cones activity cannot be excluded, considering that some cone receptive fields come surprisingly close to the smaller of the HC receptive fields, according to Nelson (1977).

We do not have direct electrophysiological evidence about the function of interplexiform cells, but neurochemical and morphological studies allow us to postulate that they build a feedback pathway, transmitting information centrifugally from the inner to the outer plexiform layers (Dowling et al. 1980).
The function of the biplexiform cell recently found in monkey retina (Mariani 1982) is still unknown.

5 Organization of the Outer Plexiform Layer in Tetrapoda Retina

The description in recent years of plexus formed by cells which have synaptic connections in the OPL and of different types of HC found in tetrapoda retina make obsolete Cajal's (1893) early assertion that in vertebrate retina "the only anatomical variations that occur are those in the relative thickness of the layers and in the form and thickness of the rods and cones." Nevertheless, Cajal's contribution to the knowledge of the structure of the retina was so impressive that for more than 70 years the physiological studies were only interpreted in the context of his structural ideas, but the new structural and functional evidence accumulated in the past 20 years makes it compulsory to know the peculiarities of each retina studied with intracellular recording and in this way avoid the risk of indiscriminately extrapolating the evidence obtained to other animal species.
If we compare mammal and bird retinas or even in mammals, primate and subprimate retinas, relevant differences in the organization of the OPL are found.

5.1 Primates

In monkey retina the cells which make synaptic connections at the OPL level are: photoreceptors (cones and rods), bipolar cells ("midget," "flat," and rod bipolars), type I and II HC, and interplexiform cells (Fig. 8).
The cone synaptic bodies are interconnected through gap junctions, building a plexus, but in primate retina the two HC types morphologically appear as individual neurons without specialized membrane contacts between them; also, physiological evidence of plexus formed by HC is lacking. The dopaminergic interplexiform cell, being presynaptic to HC (type I or II?) in the *Cebus* monkey retina, participates in the synaptic organization of the OPL.
The axonless HC, which form a plexus in subprimate mammals, have not been found in primate retina. In Golgi-stained, whole flat-mounted retinas a cell has been found whick looks like an axonless HC (Gallego 1971), but in our opinion (Gallego 1976b), as well as in Boycott's and Hopkins (1981), it can be identified as a microglia cell whose processes spread horizontally at the OPL. This cell does not show the thin terminal fibers which end in clusters of knobs, characteristic of the axonless HC, as described in low mammals, birds, reptiles, and amphibians; instead, some short and thick sproutings of the processes can be seen. The study of these cells with the Golgi-EM technique showed that these sproutings never penetrate the photoreceptor synaptic bodies.

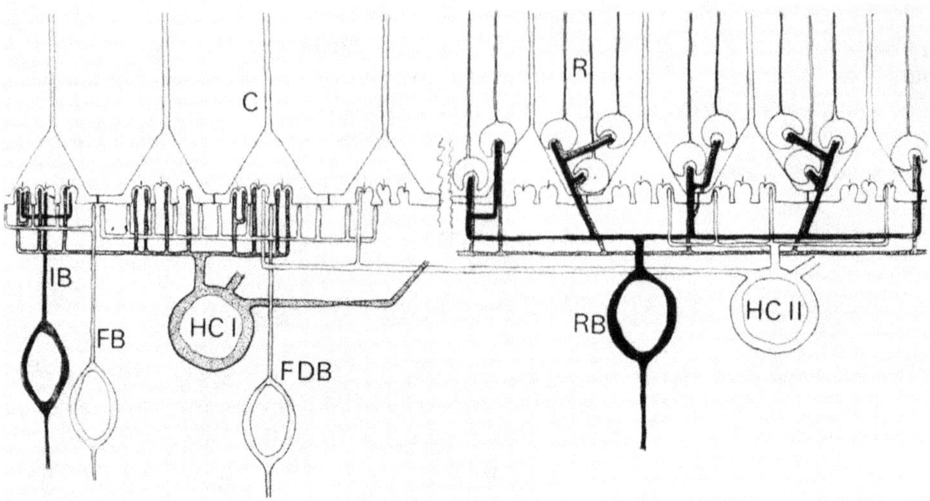

Fig. 8. Summary diagram of the outer plexiform layer of primate retina: *C*, cones; *R*, rods; *IB*, invaginating bipolar; *FB*, flat bipolar; *FDB*, flat diffuse bipolar; *RB*, rod bipolar; *HCI*, short axon horizontal cell type I; *HCII*, short axon horizontal cell type II

Type I and type II HC of the primate retina have been already described. The bipolar cells of the monkey retina, according to Kolb (1970), belong to three categories: the invaginating midget bipolar, the flat bipolars, and the rod bipolar. The invaginating midget bipolar has been found to fit a dendritic terminal process into every invagination in the cone pedicle base. The flat midget bipolar has dendritic terminals which make superficial contacts on the cone pedicle base. There are two midget bipolars per cone, flat and invaginating. The flat diffuse bipolar makes superficial contact with the cone pedicle base of about six cones. Rod bipolars contact exclusively rods, and their dendritic terminals always end as one of the central processes that penetrate the rod synaptic complex.

5.2 Low Mammals

The OPL components of the low mammals retina are: photoreceptors (cones and rods), bipolar cells (rod bipolar and two types of cone bipolar), short axon HC, and axonless HC (Fig. 9).

As in primate retina, the cone synaptic bodies are interconnected through gap junctions, but the low mammals OPL shows two main differences when compared with the OPL of the monkey retina: the plexus formed by the axonless HC and diverse types of bipolar cells.

The short axon HC and the axonless HC have already been discussed. The low mammal bipolar cells were initially described by Cajal (1893) and recently studied in the cat retina (Boycott and Kolb 1973). Cajal described one type of rod bipolar and two types of cone bipolars "with horizontally disposed dendrites destined for cones" and "giant bipolars." Eighty years later, Boycott and Kolb

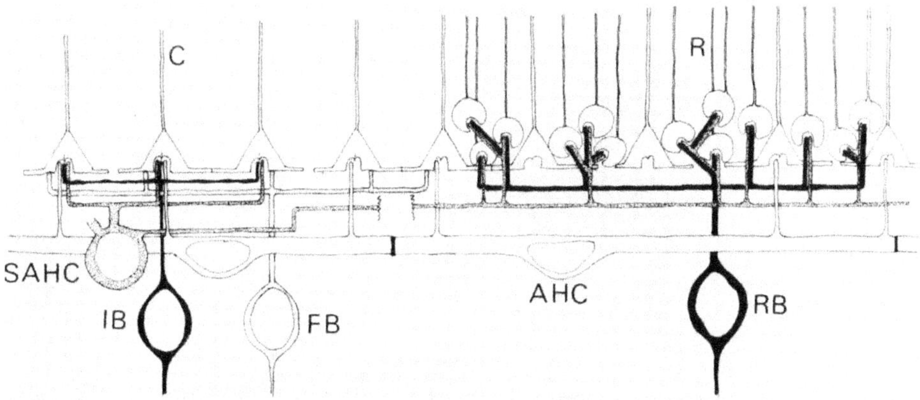

Fig. 9. Summary diagram of the outer plexiform layer of a low mammal retina: *C*, cones; *R*, rods; *IB*, invaginating bipolar; *FB*, flat bipolar; *RB*, rod bipolar; *SAHC*, short axon horizontal cell; *AHC*, axonless horizontal cell

(1973) studied the connections between bipolar cells and photoreceptors in the cat retina with the Golgi-EM technique and described two types of cone bipolars and one type of rod bipolar. The rod bipolar, according to Cajal, "has a dendritic field so rich in ascendent processes that it makes contact with 15 or 20 rod spherules." The number of rods connected to a single rod bipolar was not determined in modern studies, but it was demonstrated that "the fine dendritic terminals of the rod bipolars penetrate into the rod spherules to occupy the position of the processes opposite a synaptic ribbon" (Boycott and Kolb 1973).

In contrast with the primate retina, no midget bipolar was found in the peripheral cat retina. The two types of cone bipolars described by Boycott and Kolb were named invaginating and flat. Probably the invaginating cone bipolar is Cajal's cone bipolar, and the flat one could be identified with Cajal's giant bipolar. The invaginating cone bipolar dendrites penetrate into the triads, forming the central component of at least four to five cone synaptic bodies. The dendrites of the flat bipolar do not invaginate into the cone triads but make superficial contacts with the base of the synaptic bodies of 8 – 14 cones.

5.3 Birds

The organization of the OPL of the bird retina has not been fully clarified because, although the photoreceptors and HC have been widely studied, as far as the bipolar cell is concerned, we have to relay mainly on Cajal's study which appears to be insufficient when the different bipolar types are studied with some of the modern techniques (Fig. 10). Furthermore, a distinction between the organization of the OPL in diurnal and nocturnal birds has to be made: diurnal birds show a stratified OPL due to the distribution at three levels of the synaptic bodies of photoreceptors, while the nocturnal birds, which have fewer types of photoreceptors, have their synaptic bodies lying at a single level.

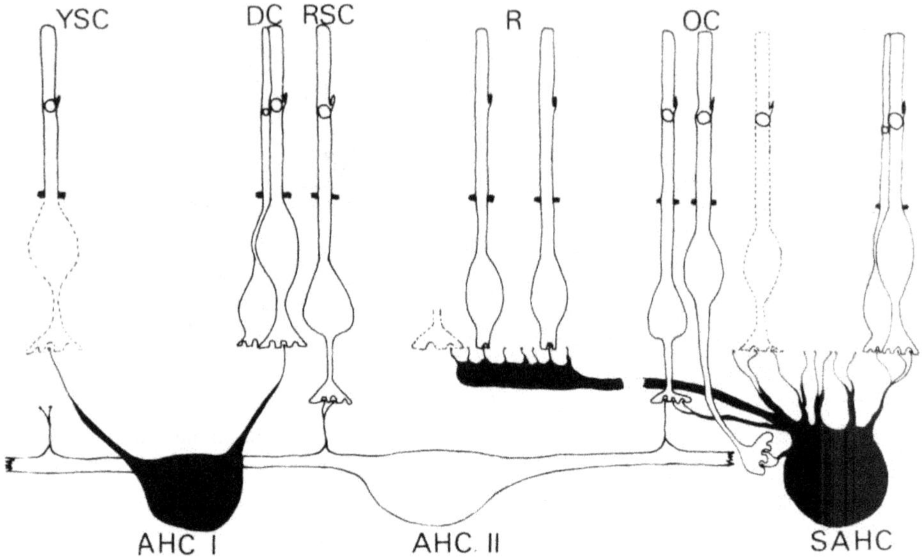

Fig. 10. Summary diagram of the outer plexiform layer of avian retina: *DC*, double cone; *YSC*, yellow single cone; *OC*, oblique cone; *RSC*, red single cone; *R*, rod; *SAHC*, short axon horizontal cell; *AHCI*, axonless horizontal cell type I; *AHCII*, axonless horizontal cell type II

In diurnal birds the cells whose processes make synaptic contacts in the OPL are: five types of photoreceptors, one type of short axon HC, two types of axonless HC, and an undetermined number of types of bipolar cells. Several plexus formed by these cells have been revealed and are formed by the cone synaptic bodies of the external level, by the axon terminals of the short axon HC, and probably by the two types of axonless HC.

With the help of a technique of enzymatic digestion of the retina and isolation of cells (Tarrés 1982), five types of photoreceptors and the color of the oil droplet they contain have been described in the chicken retina: double cones, whose main component has a green oil droplet and not yellow as stated before by several authors including ourselves; single cones, corresponding to the Cajal straight cones, of two types: one with a red oil droplet and the other with a droplet; one type of oblique cone with a pale green oil droplet; and rods.

From Cajal's work we know that the photoreceptor synaptic bodies lie at three levels in the OPL of diurnal birds (an observation confirmed by electron-microscopy studies), but only at one level in nocturnal birds (Gallego et al. 1975a). In diurnal birds the external level is formed by the synaptic bodies of rods, double cones, and perhaps a single cone; the intermediate level is formed by the synaptic bodies of Cajal's straight cones, which have a red oil droplet; and the internal level, or vitreal, is formed by the synaptic bodies of Cajal's oblique cones. The location of the synaptic bodies of the cones bearing a yellow oil droplet is dubious; they lie either at the outer or intermediate level.

The three types of HC found in diurnal bird retinas and their connections with the photoreceptors have already been discussed. Up to now we have not been able to find axonless HC in the nocturnal birds, but their presence can not be excluded. The morphology of the short axon HC of the owl retina, nearly identical to the same cell type in mammals, makes the study of the nocturnal birds retina an interesting tool to analyze the influence of the photic environment on the organization of the OPL in animals living in different ambients.

5.4 Reptiles

The structure of the OPL of diurnal birds and the turtle is similar, but while intracellular recordings of birds HC have not been reported, a good correlation has been established between HC and structural and physiological characteristics of photoreceptors (Leeper 1978a, b).

Turtle retina has five types of photoreceptors (three classes of single cones, double cones, and rods), one type of short axon HC, and two types of axonless HC. Intracellular recording has differentiated (Baylor and Hodgkin 1973) three types of single cones with spectral sensitivity peaks of 630, 550, and 460 μm and also double cones which are pairs of red- and green-sensitive elements.

From the physiological approach the short axon HC, as discussed previously, is considered as two functionally different structures: L1 HC and L2 HC. The soma of the short axon HC (L2 HC) connects with red- and green-sensitive single cones and with red-green-sensitive double cones. The axon terminal contacts red single cones, double cones, and rods. The axonless HC have no connections with rod synaptic bodies. Subtype H_2 from Leeper (1978a), which gives chromaticity R/G responses, contacts green- and blue-sensitive single cones but not double or red single cones. The connections of the axonless H_3 subtype from Leeper, which gives chromaticity G/B responses, have not been established up to now.

6 Concluding Remarks

In fish and tetrapoda retina the structural and physiological studies of individual cells with synapses in the OPL agree on the types of responses generated by light stimuli and on the general organization of their receptive fields but significant differences among species are found in cell types and their synaptic contacts. The main differences appear in the types of photoreceptors and in the organization of the lateral association system formed by HC and are evident when comparing, for instance, the fish and mammal retina, the cat and monkey retina, and the diurnal and nocturnal bird retina.

The existence in tetrapoda retina of two morphologically and probably functionally distinct HC, axonless HC and short axon HC, must be emphasized. All kinds of HC in fish retina form plexus, but in tetrapoda retina plexus are built only by the axonless HC. In submammalian species they are built only by the axon terminals of the short axon HC.

We can postulate that the short axon HC have evolved in the process of evolution from fish retina, where their somas form a plexus and the axon ends in an elongated structure with no apparent contact with photoreceptors, to the primate retina where this HC appears as a typical, short neuron which does not form a plexus and connects a small number of cone-cone bipolar synapses with a larger number of rod-rod bipolar synapses located at a distance of 2 mm. The spread of S potentials through the axonless HC plexus found in fish and cat retina is not to be expected between the short axon cells of the primate retina. As this retina has no axonless HC, the lateral interaction at the OPL level can only occur between cone synaptic bodies. Physiological evidence shows that in turtle and cat retinas, the axons do not transmit signals between the soma and axon terminals, and as a consequence, these two structures have been considered as two different HC. In primate retina, evidence of a similar characteristic is lacking, and the possibility of transmission through conducted signals always exists.

Cat retina, a very useful tool for the study of individual cells in mammals, is not a good model to analyze the function of the OPL in the primate retina. A striking difference between the two retinas is the existence of a very well-defined plexus of axonless HC in the cat, which is also found in all subprimate retinas studied. Lateral spread of activity through this plexus has been confirmed by intracellular recording.

The submammalian species also have their own characteristics. Amphibians have a short axon HC with an axon terminal morphologically similar to the same structure in fish retina, with the difference that it is located at the OPL level and like all tetrapoda retina, primates excluded, has axonless HC. Information about the connections of both HC in amphibians is lacking. Reptiles and diurnal birds have a clear difference in the organization of their OPL compared with mammals: the axon terminals build a plexus located between the somas of the short axon HC and the synaptic bodies of the receptors, and they probably have two types of HC forming plexus instead of the single one found in low mammals retina. Turtle retina studies have been very useful for obtaining information on the processing of signals generated by photoreceptors, but the evidence of great value accumulated in recent years cannot be extrapolated to primates. For instance, the axon terminals of the short axon HC are quite different in both retinas: turtle axon terminals are connected with double and red cones and with rods, while in the monkey and the cat they are connected only with rods. The large receptive field of the axon terminal, larger than the area covered by this structures, is clearly due to lateral spreading of activity through the plexus they build. As commented above, a similar finding in the primate short axon HC terminals is not expected.

HC are indeed important elements in the information processing in the vertebrate retina. Axonless HC plexus, through feedback synapses to photoreceptors, form the surroundings of the bipolar concentric receptive fields and start the processing of chromatic information. The short axon HC in a mixed rod-cone retina, as in mammals, could function as an inhibitory interneuron, suppressing rod activity at photopic stimulus intensities.

The function of the interplexiform cells as a feedback pathway from the inner to the outer plexiform layers is being studied at the present time.

Further advances in the knowledge of the mechanisms involved in processing information in the retina are expected in a near future through the use of combined morphological, electrophysiological, and, especially, neurochemical techniques applied to the comparative study of retinas which show differences in their cell types and synaptic organization.

References

Balbuena FF (1936) Conexiones de los conos y bastones al nivel de la plexiforme externa. Arch Oftalmol Hisp-Amer 35:337–363

Baron M, Gallego A (1982) The horizontal cells of the *falconidae* retina. In: Hollyfied JG (ed) Structure of the eye, 4th edn. Elsevier, pp 165–173

Baylor DA (1974) Lateral interaction between vertebrate photoreceptors. Fed Proc 33:1074–1077

Baylor DA, Hodgkin AL (1973) Detection and resolution of visual stimuli by turtle photoreceptors. J Physiol 234:163–198

Baylor DA, Fuortes MGF, O'Bryan PM (1971) Receptive fields of cones in the retina of the turtle. J Physiol 214:265–294

Boycott BB, Hopkins JM (1981) Microglia in the retina of monkey and other mammals retina, its distinction from other types of glia and horizontal cells. Neuroscience 6:679–688

Boycott BB, Kolb H (1973) The horizontal cells of the rhesus monkey retina. J Comp Neurol 148:115–140

Boycott BB, Dowling JE, Fisher SK, Kolb H, Laties AM (1975) Interplexiform cells of the mammalian retina and their comparison with catecholamine-containing retinal cells. Proc R Soc Lond [Biol] 191:353–368

Brown KT, Murakami M (1968) Rapid effects of light and dark adaptation upon the receptive field organization of S potentials and late receptor potentials. Vision Res 8:1145–1171

Cajal SR (1888) Estructura de la retina de las aves. Rev Trim Histol Norm Patol 2

Cajal SR (1893) La rétine des vertebrés. La Cellule 9–119

Cajal SR (1933) Los problemas histofisiológicos de la retina. XIV Concilium Optalmologicum Madrid, pp 1–19

Cajal SR (1904) Textura del sistema nervioso del hombre y de los vertebrados. Moya, Madrid, p 595

Copenhagen DR, Owen WG (1976) Functional characteristics of lateral interactions between rods in the retina of the snapping turtle. J Physiol 259:251–282

Crescitelli F (1972) The visual cells and visual pigments of the vertebrate eye. Springer, Berlin Heidelberg New York pp 245–363 (Handbook of sensory physiology, vol 7/1.)

Custer NV (1973) Structurally specialized contacts between the photoreceptors of the retina of the axolote. J Comp Neurol 151:35–56

De Robertis E, Franchi CM (1956) Electron microscopy observations on synaptic vesicles in synapses of the retinal rods and cones. J Biophys Biochem Cytol 2:307–318

Dowling JE, Ehinger B (1975) Synaptic organization of the amine-containing interplexiform cells of the goldfish and *Cebus* monkey retinas. Science 188:270–273

Dowling JE, Brown JE, Major D (1966) Synapses of horizontal cells in rabbit and cat retinas. Science 153:1639–1641

Dowling JE, Ehinger B, Floren I (1980) Fluorescence and electron microscopical observations of the amine-accumulating neurons of the *Cebus* monkey retina. J Comp Neurol 192:665–685

Ehinger B (1966) Adrenergic retinal neurons. Z Zellforsch Mikrosk Anat 71:146–152

Ehinger B, Falk B (1969) Morphological and pharmacohistochemical characteristics of adrenergic retinal neurons of some mammals. Albrecht von Graefes Arch Klin Exp Ophthalmol 178:295–305

Ehinger B, Floren I (1976) Indoleamine-accumulating neurons in the retina of rabbit, cat and goldfish. Cell Tissue Res 175:37–48

Fain GL (1975) Interactions of rod and cone signals in the mud puppy retina. J Physiol 252:735–769

Fain GL, Gold GH, Dowling JE (1976) Receptor coupling in the toad retina. Cold Spring Harbor Symp Quant Biol 15:547 – 561
Fisher SK, Boycott BB (1974) Synaptic connections made by horizontal cells within the outer plexiform layer of the retina of the cat and the rabbit. Proc R Soc Lond [Biol] 186:317 – 331
Foerster MH, Van De Grind WA, Grusser OJ (1977) The response of cat horizontal cells to flicker stimuli of different area, intensity and frequency. Exp Brain Res 29:367 – 385
Fuortes MGF (1972) Responses of cones and horizontal cells in the retina of the turtle. Invest Ophthalmol Vis Sci 11:275 – 287
Fuortes MG, Simon EJ (1974) Interactions leading to horizontal cell responses in the turtle retina. J Physiol 240:177 – 198
Gallego A (1953) Procedimiento de impregnación argéntina de la retina entera. An Inst Farmacol Esp 2:171 – 176
Gallego A (1964) Description d'une nouvelle couche cellulaire dans la rétine des mammiferes et son rôle fonctionnel possible. Bull Assoc Anat (Nancy) 49:624 – 631
Gallego A (1965) Connexions transversales au niveau des couches plexiformes de la rétine. Actual Neurophysiol 6:5 – 27
Gallego A (1971) Horizontal and amacrine cells in the mammals retina. Vision Res 11:33 – 50
Gallego A (1975) Las células horizontales de la retina de los vertebrados. Real Academia Nacional de Medicina, Instituto de España, Madrid
Gallego A (1976a) Comparative study of the horizontal cells in the vertebrate retina: mammals and birds. In: Zettler F, Weiler R (eds) Neural principles in vision. Springer, Berlin Heidelberg New York pp 26 – 62
Gallego A (1976b) La microglia de la retina. An R Acad Nac Med (Madr) 93:2 – 17
Gallego (1977) Participación de las células horizontales amacrinas en el procesamiento de la información visual. Neurobiología 25:41 – 54
Gallego A (1978) Mecanismos neurales de adaptación visual a nivel de la capa plexiforme externa de la retina. Fundación Juan March, Serie Universitaria 52, Madrid
Gallego A (1982) Organization of the outer plexiform layer of the tetrapoda retina: horizontal cells of mammals and avian retina. In: Hollyfield JG (ed) Structure of the eye, vol 4. Elsevier, pp 151 – 164
Gallego A, Perez Arroyo M (1976) Photoreceptors and horizontal cells of the turtle retina. In: Yamada, Mishima (eds) The structure of the eye III. 5:311 – 317
Gallego A, Sobrino JA (1975) Horizontal cells of the monkey retina. Vision Res 15:747 – 748
Gallego A, Baron M, Gayoso M (1975a) Organization of the outer plexiform layer of the diurnal and nocturnal bird retina. Vision Res 15:1027 – 1028
Gallego A, Baron M, Gayoso M (1975b) Horizontal cells of the avian retina. Vision Res 15:1029 – 1030
Genis-Galvez JM, Prada F, Armengol JA (1979) Evidence of three types of horizontal cells in the chick retina. Jpn J Ophthalmol 23:378 – 387
Grüsser OJ (1960) Receptorabhängige Potentiale der Katzenretina und ihre Reaktion auf Flimmerlicht. Pflügers Arch 217:511 – 525
Kaneko A (1970) Physiological and morphological identification of horizontal, bipolar and amacrine cells in goldfish retina. J Physiol 207:623 – 633
Kaneko A (1971) Electrical connections between horizontal cells in the dogfish retina. J Physiol 213:95 – 105
Kolb H (1970) Organization of the outer plexiform layer of the primate retina: electron microscopy of Golgi impregnated cells. Philos Trans R Soc London [Biol] 258:261 – 283
Kolb H (1974) The connections between horizontal cells and photoreceptors in the retina of the cat: electron microscopy of Golgi preparations. J Comp Neurol 155:1 – 14
Kolb H (1977) The organization of the outer plexiform layer in the retina of the cat: electron microscopy observations. J Neurocytol 6:131 – 153

Kolb H, Mariani A, Gallego A (1980) A second type of horizontal cell in the monkey retina. J Comp Neurol 189:31 – 39
Lamb TD, Simon EJ (1976) The relation between intercellular coupling and electrical noise in turtle photoreceptors. J Physiol 262:257 – 286
Leeper HF (1978a) Horizontal cells of the turtle retina I. Light microscopy of Golgi preparations. J Comp Neurol 182:777 – 794
Leeper HF (1978b) Horizontal cells of the turtle retina II. Analysis of interconnections between photoreceptor cells and horizontal cells by light microscopy. J Comp Neurol 182:795 – 809
Leeper HF, Copenhagen DR (1979) Mixed rod-cone responses in horizontal cells of snapping turtle retina. Vision Res 19:407 – 412
Marc RE, Lam DMK (1981) Glycinergic pathways in the goldfish retina. J Neurosci 1:152 – 165
Mariani AP (1982) Biplexiform cells: Ganglion cells of the primate retina that contact photoreceptors. Science 216:1134 – 1136
Mariani AP, Leure-Dupree AE (1977) Horizontal cells of the pigeon retina. J Comp Neurol 175:13 – 26
Miller WH, Hashimoto Y, Saito T, Tohita T (1973) Physiological and morphological identification of L- and C-type S potentials in the turtle retina. Vision Res 13:443 – 447
Missotten L (1965) The synapses in the human retina. In: Rohen SW (ed) Structure of the eye II, Symposium Schattauer, Stuttgart, pp 17 – 28
Missotten L, Appelmans M, Michiels J (1963) L'ultra-structure des synapses des cellules visuelles de la rétine humaine. Bull Mem Soc Fr Ophtalmol 76:59 – 82
Mitarai G, Svaetichin G, Vallecalle E, Fatehchand R, Villegas J, Laufer M (1961) Glia-neuron interactions and adaptational mechanisms of the retina. In: Jung R, Kormhuber H (eds) The visual system: Neurophysiology and psychophysics. Springer, Berlin Heidelberg New York, pp 463 – 481
Molenaar J (1981) Intracellular light responses in the intact in vivo eye of the cat. PhD dissertation, University of Amsterdam, The Netherlands
Naka KI, Rushton WAH (1967) The generation and spread of S potentials in fish *(Cyprinidae)*. J Physiol 192:437 – 461
Negishi K, Drujan BD (1978) Effects of catecholamines on the horizontal cell membrane potential in the fish retina. Sens Processes 2:388 – 395
Nelson R (1977) Cat cones have rod input: A comparison of the response properties of cones and horizontal cell bodies in the retina of the cat. J Comp Neurol 172:109 – 136
Nelson R, Lutzow Av, Kolb H, Gouras P (1975) Horizontal cells in cat retina with independent dendritic systems. Science 189:137 – 139
Nelson R, Kolb H, Famiglieti EV, Gouras P (1976) Neural responses in the rod and cone systems of the cat retina: Intracellular records and procion stains. Invest Ophthalmol Vis Sci 15:946 – 953
Niemeyer G (1975) The function of the retina in the perfused eye. Doc Ophthalmol 39:53 – 116
Norman RA, Kolb H (1981) Anatomy and physiology of the horizontal cells of the visual streak region of the turtle retina. Vision Res 21:1585 – 1588
Orellana JM, Gallego A (1959) Distribución topográfica de las células horizontales de la retina. Actas Soc Esp Cienc Fisiol 5:251 – 252
Piccolino M, Neyton J, Witkovsky P, Gerschenfeld HM (1982) GABA antagonists decrease functional communication between L-horizontal cells of turtle retina. Neruosci Lett [Suppl] 10:387 – 388
Polyak S (1941) The retina. University of Chicago Press, Chicago
Raviola E, Gilula NB (1973) Gap junctions between photoreceptor cells in the vertebrate retina. Proc Natl Acad Sci USA 70:1677 – 1681
Simon EJ (1973) Two types of luminosity horizontal cells in the retina of the turtle. J Physiol 230:199 – 211

Simon EJ (1974) Feedback loop between cones and horizontal cells in the turtle retina. Fed Proc 33:1078 – 1082

Simon EJ, Lamb TD, Hodgkin AL (1975) Spontaneous voltage fluctuations in retinal cones and bipolar cells. Nature 256:661 – 662

Sjöstrand FS (1953) The ultrastructure of the inner segments of the retinal rods of the guinea pig eye as revealed by electron microscopy. J Cell Comp Physiol 42:45 – 70

Sjöstrand FS (1958) Ultrastructure of retinal rod synapses of the guinea pig eye as revealed by three-dimensional reconstructions from serial sections. J Ultrastruct Res 2:122 – 170

Sobrino JA, Gallego A (1970) Células amacrinas de la capa plexiforme externa de la retina. Actas Soc Esp Cienc Fisiol 12:373 – 375

Steinberg RH (1969) Rod-cone interaction in S potentials from the cat retina. Vision Res 9:1331 – 1344

Steinberg RH (1971) Incremental responses to light recorded from epithelial cells and horizontal cells of the cat retina. J Physiol 217:93 – 110

Steinberg RH, Schmidt R (1970) Identification of horizontal cells as S-potential generators in the cat retina by intracellular dye injection. Vision Res 10:817 – 820

Stell WK (1965) Correlation of retinal cytoarchitecture and ultrastructure in Golgi preparations. Anat Rec 153:389 – 397

Stell WK (1967) The structure and relationship of horizontal cells and photoreceptor bipolar synaptic complexes in goldfish retina. Am J Anat 120:401 – 423

Stephan P, Weiler R (1981) Morphology of horizontal cells in the frog retina. Cell Tissue Res 221:443 – 449

Svaetichin G (1953) The cone action potential. Acta Physiol Scand 29 [Suppl 106]:565 – 600

Svaetichin G, McNichol EF (1958) Retinal mechanisms for chromatic and achromatic vision. Ann NY Acad Sci 74:385 – 404

Svaetichin G, Negishi K, Fatehchand R, Drujan B, Selvin De Testa A (1965) Nervous function based on interactions between neuronal and non-neuronal elements. Prog Brain Res 15:243 – 266

Tarrés MAG (1982) Capa plexiforme de la retina de las aves. PhD dissertation, Universidad Complutense, Madrid

Toyoda J, Hashimoto H, Anno H, Tomita T (1970) The rod response on the frog as studied by intracellular recording. Vision Res 10:1093 – 1100

Werblin FS, Dowling JE (1969) Organization of the retina of the mud puppy, *Necturus maculosus* II. Intracellular recording. J Neurophysiol 32:339 – 355

West RW (1978) Bipolar and horizontal cells of the gray squirrel retina: Golgi morphology and receptor connections. Vision Res 18:129 – 136

Whitten DN, Brown KT (1973) Photopic suppression of monkey's rod receptor, apparently by a cone-initiated lateral inhibition. Vision Res 13:1629 – 1658

Subject Index

www.ingramcontent.com/pod-product-compliance
Lightning Source LLC
Chambersburg PA
CBHW061220240926
51846CB00006B/1289
* 9 7 8 3 6 4 2 6 9 1 6 5 2 *